George Washington Never Slept Here
The history of street names in Washington, D.C.

Amy L. Alotta

Bonus Books, Inc., Chicago

© Bonus Books, Inc., 1993
All rights reserved

Except for appropriate use in critical reviews or works of scholarship, the reproduction or use of this work in any form or by any electronic, mechanical or other means now known or hereafter invented, including photocopying and recording, and in any information storage and retrieval system is forbidden without the written permission of the publisher.

97 96 95 94 93   5 4 3 2 1

Library of Congress Catalog Card Number 92-74146

International Standard Book Number: 0-929387-82-1

Bonus Books, Inc.
160 East Illinois Street
Chicago, Illinois 60611

First Edition

Printed in the United States of America

Photographs by Amy L. Alotta unless otherwise noted.

To visions that become reality.

Contents

Foreword—vii

Preface—xi

Introduction—xv

Letter Streets—1

Named Streets—3

Bridges—151

Bibliography—161

Index—165

Foreword

For more than a quarter century, I have been bothering Amy Alotta with the stories behind the names of Philadelphia streets and other places across the nation. Turn about, I have learned, is truly fair play! It is now Amy's turn. Happily, when it is your daughter, it is no bother at all.

When Amy and her brother Peter were small children, their mother and I would take them everywhere . . . from my pursuit of military forts to my work in the warm climes of Boca Raton, Florida, or Six Flags Over Texas, in Arlington. No matter where we went, the children's curiosity was such that the old man had to check through books, etc., to find answers to their questions . . . including Peter's "Why does the Liberty Bell have a crack and a crapper?" I had to explain to him that he meant "clapper," but the word kept coming out with the "r." I also had to explain why we lived on Waverly Street. I did sluff them off a bit on that one. Later I found it was named for one of Sir Walter Scott's books.

Amy, on the other hand, was the shy one of the family. She let her brother ask the stupid questions. Little did I know then that she tucked all that information away for future use. Now it is her time to provide the answers to other people's "stupid" questions.

In our family, however, there is never such a thing as a stupid question; there are only stupid answers. Every question has an answer, but sadly, some people prefer to make up answers rather than do their homework. I can assure you, the reader, that Amy Alotta has done her homework and her comments about the street names of Washington are on target!

Foreword

Washington has always been the town where Amy wanted to live and work. When our family was helping her decide on where to go to college, there were no compromises. She had her mind set on The American University in Washington, D.C. There was never a question in Amy's mind; she was going to the District come hell or high water. When she arrived at American, my wife and I were worried about her. After all, we were living in Mississippi at the time—part of my growing-up—and were many miles away from her to provide support. But Amy was not alone. Gjore Mollenhoff, a dear friend who died of cancer at the youthful age of 48, took Amy under her wing and acted as surrogate parent. Gjore even arranged for Amy to "work" at Ronald Reagan's inauguration in 1981 and to experience the vibrant life of the nation's capital.

Amy thrived in Washington and we, her mother and father, realized we had nothing to fear; Amy took to Washington life like the proverbial duck took to water. Together with college roommate and bosom buddy Dawn Kessner Hall, Amy began her first research into the nation's capital. The two young women located all of the bathrooms in the City of Washington! She and Dawn figured that would make a great tour guide, because all they ever heard from people from out of town was "Where's the john?" Sadly, *The Washington Post* beat them to publication. The search for the "johnnies on the spot" was their first excursion into researching Washington.

After Amy graduated from American and went to work as a food service manager at The National Zoo, I began to realize that Amy had been trained too well at home. One day when she was driving her mother and me through the environs of the District, out of the blue she asked, "Dad, do you know where Old Gallows Road got its name?" Before I could answer, she went into a five-minute exposition of the name. I could not have done it better myself. My little girl was now a woman. I must say that her description of street names did not end with one!

When Aaron Cohodes, publisher of Bonus Books, and I began the "Street Names of America" project following publication of my *Mermaids, Monasteries, Cherokees, and Custer: The Stories Behind the Street Names of Philadelphia*, we thought that the nation's capital should be one of the first books to be published as part of the series. I

Foreword

scratched my head for about 30 seconds and called Amy and asked if she would feel comfortable doing such a book. She did feel a certain discomfort at first—as would any first-time author. But she did feel at ease with the subject, the research and the excitement of discovery. Her father knew she was hooked on the subject when she gave me my very own street sign—ALOTTA BOULEVARD, N.W.—at Christmas last year. There is no such place in the District, but after the publication of this book, there should be . . . and it should honor Amy Alotta.

Amy has done a great deal for the City of Washington. She has found out what few people knew before: how chaotic the naming process has been in the District. She learned, as her father before her, that bureaucrats keep few records on the naming process for fear someone will find out they were applying a name as a sop for a friend or someone with whom they wished to curry favor.

The only problem I, as general editor, had with my daughter, as writer, was that she did not want me to see the preliminary drafts before she was through with them. She wanted to pull it all together herself and then show it to me. I fear that my dominant personality made her apprehensive of what I might do to her document. Was she not surprised when I finally looked it over and did not change more than a few words, but offered an idea or two on how to pursue other angles of research? She learned that an editor is at his—or her—best when the author's words are not distorted or contorted or made to sound like they came out of the mouth—or pen—of the editor.

George Washington Never Slept Here is a great book—and not because the author is my daughter. It is a great book because much time and effort went into its research and writing. I enjoyed reading the book from two points of view . . . that of a proud father, but most of all that of a proud editor. In either case, I've popped all my buttons.

From what I gather from our publisher, Amy is hard at work on her second book: a guide to the zoos in America. The child has flown the coop.

Robert I. Alotta, Ph.D.
General Editor
Street Names of America Project

Preface

The other day, while driving through Washington, I wondered how I had become interested in street names. After much consideration, I found I could not pinpoint exactly what had piqued my curiosity.

It may have begun with my father's incessant questions and detailed answers. These were directed at my brother and I while he was conducting research on Philadelphia street names. As I child, I dreaded getting into a car with him. It took approximately 5.3 minutes before we passed a street with a past. "Do you know how Arch Street got its name?" My answer was usually, "No!" I learned quickly that guessing prolonged the ordeal, often leading to stories about famous people and events unrelated to the original subject. My father told the same stories at parties. I pitied those who unknowingly presented him with a fresh ear.

I attempted to outwit him. I began to read his manuscript secretly, hoping to find the answers to his questions. Unfortunately, I was always in the wrong part of the alphabet. I knew A through D and he would ask about Girard Avenue.

As our family moved around the country, so, too, did my father's questions about street names. Eventually, he expanded his research to place names, increasing the odds that I would be unable to answer his questions.

After one particularly gruelling trip, I vowed I would never torment people with facts the way my father did so often. I tried very hard to avoid my father's keen interest in nomenclature and history, but it crept up on me.

Preface

My interest also may have stemmed from all those excruciating hours spent in history classes. I listened to at least a dozen teachers regurgitate facts from history books. I made many, often failed, attempts to memorize all the dates, people and events. I knew from my father's narrations that history was filled with intriguing and, often, humorous stories, but I rarely heard these in school.

I was sure college would be different. I realized how wrong I was while attempting to remain conscious in a class about the U.S. presidents. The professor spewed out only the well-known facts about the presidents; information anyone could have found in the encyclopedia. I swore that if I ever broke the above vow, I would tell the interesting tales of history.

Of course, not all my college professors were purveyors of useless information. **Kathleen Burns** drilled into me the importance of double and triple-checking facts. While doing research, I often found two sources that stated one origin of a street name. Kathy Burns' voice would echo in my ears. Several times I found a third source that added new light or completely refuted the first two . . . and the search would continue.

James P. Lynch, professor of justice at The American University, is my surrogate father. Whether he knows it or not, his advice has always been right on the mark. He was one of the first I told about the book. He reacted with the perfect mix of excitement, curiosity, concern and one question: "How did Belt Road get its name?" Since this was the first street anyone had asked me about, I was determined to find its origin. It took me a while to discover, but like the spaghetti sauce, it's in there.

Philip Ogilvie, curator of the D.C. Archives and Records Center, handed me research he had started on District alleys. He also showed me a progressive display of Washington maps, from the beginning of the city through the 1920s. This proved incredibly helpful, since the mounds of paper from the commissioners often presented conflicting information. **Dorothy Provine,** also of the D.C. Archives, tirelessly pulled files from the shelves and answered my endless barrage of stupid questions. I'm amazed at how much knowledge about Washington these two have stored in their brains. I hope the city knows how lucky it is to employ these two individuals.

Preface

A nameless operator at the D.C. Government switchboard destroyed my image of the telephone operator painting her nails while placing people on eternal hold. I was trying to find the number for the D.C. Archives. She read a long list of names and numbers for me to try. I told her that I had already tried some, but they had been changed or disconnected. She could have told me to try the rest, but she did not. She dialed six numbers before finding the right one. Not only was I impressed, my right index finger was extremely grateful.

There were other nameless voices, too. With sixty-seven things going at once, I forgot to ask the names of those who answered my questions at the Washington Historical Society and at the Washingtoniana Division of the Martin Luther King Library.

At one point, I became increasingly frustrated by the repetition of several phrases: "I don't know where you'd find that information," "We don't have it here," "Are you a student? We don't provide this information to just anybody," and, my personal favorite, "You want to know the history of what? What a stupid idea." Since I have the face of a 12 year old, I heard the student question a great deal. I have been out of school for six years; I deeply resented the question. I also learned that students don't get much respect in research circles. That's a shame.

At the height of my frustration, I met **Brent Berger, M.D.** He showed an interest in my research and his enthusiasm was infectious. (His choice was Jenifer Street.) In conversation, he pinpointed one reason I enjoy research. He said, "You know, one of the reasons I became a doctor was because I like to play detective. I like to solve mysteries." Although our chosen professions are quite different, I have to agree with him that it is thrilling to find that elusive piece of information that brings the puzzle together. Unfortunately, for the thrill-seeker and perfectionist in us, that information sometimes remains concealed despite our best efforts.

The staff of Bonus Books exercised patience when my real job prevented me from spending more time on the manuscript.

The manuscript was completed on MultiMate. The program has some annoying quirks, but it also has one terrific feature: it saves automatically at the end of each

Preface

page. This feature kept my blood pressure level each time Pepco failed to keep the electrical current running to my house. (This seemed to happen every time I sat down at the computer.)

My colleagues at Friends of the National Zoo and an assortment of friends and acquaintances witnessed my worst fear become reality. I now torment them regularly with the stories behind the street names, whether they want to know or not. I offer fair warning to all who enter my car. The best part, though, is that my father now knows how it feels to be on the receiving end.

Introduction

Human beings have been naming things for centuries. We name everything: places, bodies of water, mountains, buildings, and of course, streets. The names we give reflect much about society. These choices tell much about the namers. Some are drawn from ego, some from a desire to honor role models, and still others are descriptive of the area through which they travel. The street names in the District of Columbia tell much about the history of the city and the nation. Before you dive into the history of Washington street names, though, it is important to see how the city developed and how vital streets were in the original plan.

During the late 1700s, Congress moved from city to city in the Northeast. In 1783, an unruly mob of Pennsylvania militiamen gathered outside the Philadelphia statehouse, protesting the state's failure to pay them. Members of Congress asked the president of Pennsylvania to call out the Philadelphia militia to protect them. He refused. Congress, feeling betrayed by the city and fearing that the mutiny would grow, adjourned. They agreed to reconvene in Princeton, New Jersey.

Hasty departures by Congress were not unusual. Congress left Philadelphia several times during the Revolutionary War, but it always returned. The cycle remained unbroken as Congress returned to the city in 1787. This time was different, though, as talk of a permanent capital for the nation prevailed in Congressional sessions.

According to the Constitution, Congress could establish its own jurisdiction. Many cities were considered. Princeton, N.J., was too small; New York was too

Introduction

far north. The stream of suggestions continued, each followed by a rejection. The Northerners wanted the capital in the North and the Southerners favored the South.

The debate continued; an equitable decision appeared impossible. In January 1789, Alexander Hamilton proposed the Assumption Act to Congress. Under this act, the debts incurred by the states during the Revolutionary War would be assumed by the federal government. Hamilton realized that many of these debts were owed to wealthy, influential men. He hoped these men would fight for the new nation if they had a stake in its survival. Many of the war debts were incurred by Northern states. The Southern states opposed the act knowing that the funds would be raised by an increase in import tariffs. The Southern states, consisting mostly of farming communities, needed the imports to survive. Hamilton used the placement of the capital city as a bargaining chip. His deal: He would vote for placement of the capital city in the South if the Southerners would support his act. He succeeded.

Congress now needed to choose a location for the new federal city. A site along the Potomac River appeared most promising because it would allow passage to the west. Citing the weather and the swampy conditions, several congressmen stated that living in this region would be detrimental to people's health. A resident of the area countered with examples of men who had survived these "poor" conditions well into their nineties and beyond.

To entice Congress into placement of the capital in their region, Virginia and Maryland pledged $120,000 and $72,000, respectively. In July 1790, Congress decided the federal city should be placed along the Potomac River. At the same session, Congress agreed to remain in Philadelphia for ten years until their new home was completed.

As a diversion during the capital city debate, Congress hired a young Frenchman to convert the New York City Hall into a Federal Hall for congressional use. Pierre Charles L'Enfant, a major in the American Army Corps of Engineers, created a uniquely American building using the $65,000 raised by the city. Into the architecture he incorporated the stars and stripes, medallions em-

Introduction

bossed with "U.S.," and an eagle with 13 arrows (to symbolize the number of states in the Union) extending from its talons. Learning of Congress's plans to build a capital, L'Enfant wrote to Washington and offered to design the new city. Impressed by "Monsieur Lanfang's" ability to create on a large scale, Washington agreed.

Of course, the exact location for the city had to be chosen before any other plans could be made. This task was given to Washington. He was instructed to choose a ten-mile square along the 110-mile stretch of the Potomac between the Eastern Branch (Anacostia River) and the Conogocheague River. Washington, who knew the area well, visited many cities, including Cumberland and Georgetown, Maryland, and Alexandria, Virginia. Washington requested maps from each city. He required each map to show land ownership, springs, high- and low-lying areas and land susceptible to flooding. It is thought that Washington was looking for unhealthy swamps.

Washington placed the ten-mile square at the junction of the Eastern Branch and the Potomac River. The jurisdiction included the city of Alexandria and 1,200 acres of land owned by Washington. The site he chose was as close to his home, Mount Vernon, as he could get within Congress's boundaries.

Congress, when it empowered Washington to choose the federal site, also allowed him to select three commissioners to oversee the construction of federal buildings. Congress presented one stipulation: The buildings had to be constructed on the Maryland side of the river.

Washington chose Daniel Carroll, Thomas Johnson and Dr. David Stuart as his commissioners. He also requested that Major Andrew Ellicott survey the area. Ellicott, with the help of astronomer Benjamin Banneker, began surveying the 6,100 acres from Georgetown, Maryland, to the Eastern Branch in early 1791. Ellicott was not impressed with the area:

> The Country thro' which we are now cutting one of the ten-mile lines is very poor; I think for near seven miles on it there is not one House that has any floor accept the earth; and what is more strange, it is in the neighborhood of Alexandria, and George-Town.

Introduction

> This country intended for the Permanent Residence of the Congress, bears no more proportion to the country about Philadelphia and German-Town, for either wealth or fertility, than a crane does to a stall-fed Ox!

Meanwhile, L'Enfant worked feverishly and produced, in approximately three months, an incomplete plan for the city. On June 28, 1791, while riding through the area with L'Enfant, George Washington approved the plan. Using the city of Versailles as a guide, L'Enfant placed the Capitol facing east with three wide avenues radiating from its facade. He proposed placing the President's House on Jenkins Hill, which he proclaimed "a pedestal waiting for a monument." His design also included a canal through the center of the city and a "Grand Avenue," lined with elegant buildings and shade trees. L'Enfant believed this expanse, now the Mall, would serve as a prime commercial and recreational area.

The streets were an integral part of L'Enfant's grand design. He laid the streets in a grid pattern and overlapped them with wide diagonal avenues. According to his plan, the streets were to be 80 feet wide and the avenues 160 feet wide. The avenues provided direct routes between important points in the city. The Capitol and the President's House were joined by what L'Enfant deemed the most important avenue in the city (Pennsylvania Avenue). The confluence of the streets and avenues formed circles. L'Enfant's design also included 13 squares, which represented each state in the Union.

L'Enfant was proud of his plan and did not want it altered in any way. The commissioners soon discovered that the artist had a fiery temper and was unable to take orders. Daniel Carroll, nephew of the commissioner with the same name, began construction of his house in the middle of one of L'Enfant's proposed avenues (New Jersey Avenue). L'Enfant flew into a rage. Stating that Carroll was not giving his plan proper respect, he tore down the walls of the house, ignoring orders from George Washington and Thomas Jefferson to cease.

The commissioners grew increasingly agitated over L'Enfant's antics. On September 8, 1791, the commissioners met without him. The primary purpose for the meeting was to discuss the engraving of the city map. L'Enfant wanted the map to be an exact duplicate of his plan, but the commissioners had other ideas. They decided that only two squares, the president's square and Capitol Square, should be included on the engraving. This meeting also gave birth to the city's name, since the engraving required one. The district was named Territory of Columbia; the city within was dubbed Washington, to honor the first president. The commissioners also determined street names: North/south streets would carry numbers and east/west streets would be designated with letters.

Washington asked L'Enfant to oversee the engraving of the map, but the version he gave to the engraver was unacceptable. L'Enfant had become so involved with the sale of lots that he did not have the time, nor the ambition, to prepare the map correctly. Washington requested that Andrew Ellicott complete the map. Upon seeing the completed map, L'Enfant flew into one of his now-famous rages. He claimed that his plan had been misinterpreted and his vision destroyed. Citing his failure to comply with the commissioners' decisions, Washington, Jefferson and Alexander Hamilton dismissed L'Enfant. Despite L'Enfant's dismissal, Washington wanted L'Enfant's name to appear on the engraving, but only Ellicott's name appeared on the final version.

Despite a lack of funds, the city began to develop. Buildings were built and streets were graded. Since the names of the grid streets had already been determined, the only decision that remained concerned naming the diagonals. L'Enfant's plan designated the diagonals "avenues" and Thomas Jefferson gave them the names of the 13 original states, with respect to geographical location. Jefferson played a little favoritism, though. After discovering that Virginia Avenue was shorter than the other diagonals, Jefferson altered L'Enfant's plan by extending the street across the city. Incidentally, this designation of "avenue" was the first use of the term in an address.

In 1800, as planned, Washington was completed and the government moved in. George Washington, though, never had the chance to see the city he helped create. He died on December 14, 1799.

At first, it looked like the city was going to fail. Lots were not selling and citizens seemed unwilling to take up long-term residence. The city planners seemed to be placing their emphasis on monuments instead of func-

Introduction

tional buildings. Poet Thomas Moore penned his description of the capital city:

> . . . this Embryo Capital, where fancy sees,
> Squares in morasses, obelisks in trees,
> Which second-sighted seers, even now adorn
> With shrines unbuilt and heroes yet unborn.

The city of Philadelphia attempted to woo the government back. Despite their efforts, it became evident that the government would stay in Washington. Once the government committed, the city began to grow. Within a short period, the city was larger than L'Enfant's plan.

As the city became more populated, people built their residences farther and farther from the center of the city and the government buildings. Farms and mills sprouted up along the rivers and Rock Creek. People began cutting roads through the countryside to gain direct access to the city from their "suburban" properties. The property owners named these roads for themselves (Tayloe's Road), their property (Harewood Road), their loved ones or whatever else struck their fancy. By the mid-1800s, the street system was a mess. Streets in different sections of the city bore the same names; several streets carried different names for each block. Efficient travel and mail delivery were extremely difficult.

The Civil War compounded this mess. The city had been neglected. The weight of the equipment carried through the city had rutted the streets, the parks had become overgrown and the Washington Canal was a health hazard. Horace Greeley described the city in the *New York Tribune,* July 13, 1865: "Washington is not a nice place to live in. The rents are high, the food is bad, the dust is disgusting, the mud is deep, and the morals are deplorable."

By 1869, a group of Southern and Western representatives began a push to have the capital relocated. In 1871, Ulysses S. Grant signed a bill that created a territorial form of government for the District. In an attempt to appease the citizens and federal government officials, the government promised mass city improvements. Alexander Robey Shepherd served as Vice-Chairman of the Board of Public Works. He supervised many of the improvement efforts, which included installing street lights, digging sewers and grading and paving streets.

Introduction

In 1873, Shepherd became governor of the District. Shepherd continued his plans for improvement, but his high living brought suspicion.

Shepherd's administration came under attack for corruption. His improvements placed the city on the brink of bankruptcy. The accusations were never substantiated, but Shepherd's spending forced the abolition of the territorial government in 1874. Shepherd fled the city to escape the controversy. Despite the accusations, though, Shepherd did much to improve Washington.

The improvements continued. In 1882, a project to resurrect Potomac Flats (land on which the Jefferson Memorial now sits) began. Major General Peter C. Hains of the Army Corps of Engineers supervised the project. It took eight years to reclaim land, which would be used to expand the city center, from the river.

In 1880, Georgetown became part of the District of Columbia. The city was renamed West Washington. The streets in this new section were renamed to conform to the Washington plan. The plan, though, was becoming more confused. On December 6, 1899, Congress initiated a system for naming streets in the District of Columbia. They authorized the board of city commissioners to rename existing streets that did not comply with the designated nomenclature. Congress applied the following rules that still govern street naming in the District of Columbia:

1. **North/south streets** are numbered consecutively in each direction, commencing at the meridian of the Capitol.
2. **East/west streets** are lettered, then given one-, two- and three-syllable names in alphabetical order. The names should be those of distinguished Americans in this order: Presidents, Vice Presidents, Chief Justices, Speakers of the House, distinguished members of the cabinet and other distinguished American citizens. The streets that follow should be named for members of the plant kingdom in alphabetical order.
3. **Diagonals** are named for states and territories of the United States.
4. **Streets not part of the rectangular plan** are designated roads, drives, places, courts, etc., and bear the names of a prominent local feature. Other names may be approved by the government.
5. **Circles** are named after individuals who have been prominent in their service to the country.

Introduction

6. **Alleys** are to remain nameless unless they provide the only access to a residential or commercial property.
7. Streets that are aligned bear the same name.
8. Streets that travel in the same general direction of a grid street are to carry the name of the next principal street nearest the Capitol.

Congress appropriated $3,000 to change the existing street signs. Incidentally, Washington's street nomenclature is virtually unique; two states, Illinois and Iowa, have cities that follow a similar street naming plan.

The task of renaming the streets proved monumental. Residents expressed a range of opinions to the commissioners. Some wanted no change, some favored the changes and others wanted the option to propose names (usually their own or that of a relative). The commissioners filed each request and responded to many, patiently explaining the guidelines for the street nomenclature.

The reorganization was completed in 1906. When the task was complete, very few of the original street names remained. Those that still exist today are Dumbarton, Prospect, Olive, Grace and Potomac. The confusion did not cease, though. The January edition of the Chesapeake and Potomac Telephone Directory listed the old street names.

Improvement of the streets, and the surrounding city, continued. The improvements grew from necessity; vehicular travel was becoming more popular and the streets more congested and dangerous. In 1925, electric traffic lights were installed around the city. These lights replaced semaphore signals.

Conservation efforts were gaining support, too. In the late 1800s, the suggestion was made to create a park within the city that would be untouched by commercial and residential development. Debate raged and in 1924, the wheels of bureaucracy began to grind toward the creation of Rock Creek Park. In 1927, Congress approved the creation of the National Arboretum. The idea had been proposed by Secretary of Agriculture James Wilson 28 years earlier. The arboretum, he stated, would serve as a "place for all trees that will grow in Washington, D.C."

Development and redevelopment continued. Bridges were constructed to handle the increasing traffic in and out of the city. Highways were cut to carry larger numbers of vehicles. Small streets that had had little purpose were widened. The District's unique bureaucracy hindered organized development in the city. During its brief history, Washington has changed from federal control to home rule a number of times. When it appeared the city could not handle its problems, Congress stepped in.

This constant power struggle is evident when conducting research. Since the bodies governing street naming changed with the government of the city, the records are scattered. Needless to say, no one kept great records, and if they did, they have long since been shuffled into the information netherworld.

At the District of Columbia Archives and Record Center, I found the Engineering Department records. These records detailed the street reorganization of 1905, including thousands of citizen complaints and responses from the commissioners. Many of the letters were written in flowery script that, when the ink ran or the author saved paper by writing on both sides, became nearly impossible to decipher. Some of the responses were recorded on onion skin. I learned that onion skin becomes very flimsy with age and tends to curl . . . a lot. I often felt I spent more time straightening documents than I did reading them.

The National Archives provided me with records from the commissioners and other governing bodies, as did the Washingtoniana Division of the Martin Luther King Library.

At the Historical Society of Washington I discovered that several historians had grazed the surface investigating the naming of Washington streets, but I was disappointed to find that some succumbed to retelling popular legends.

Legends, I soon discovered, dominate many works. I found many authors who stated that the omission of J Street was a slight to Chief Justice John Jay. In reality, the reason is quite simple. The cursive I and J are similar and the city fathers wanted to avoid confusion.

The history of street names is intriguing. The stories we fabricate to explain them are almost as interesting as the facts. The street names of Washington tell much about the city's founders, the people who made the capital their home and those who have passed through. All have left their mark. I hope you will enjoy your journey.

Letter Streets

As part of the original plan for street nomenclature in the District of Columbia, lettered streets advance north and south from the Capitol. Some letters do not exist in the street system: J, X, Y and Z. As noted in the Introduction, a popular story contends that J Street was omitted as a slight to Chief Justice John Jay; the existence of Jay Street proves this theory wrong. More likely, the street was omitted to avoid confusion with I Street, since cursive I's and J's are formed similarly.

A Street (NE, SE)
B Street (SE)
C Street (NE, NW, SE, SW)
D Street (NE, NW, SE, SW)
E Street (NE, NW, SE, SW)
F Street (NE, NW, SE), Street Terrace (SE)
G Place (NE), Street (NE, NW, SE, SW)
H Place (NE), Street (NE, NW, SE, SW)
I Street (NE, NW, SE, SW)
>   Many visitors to Washington become confused when they see addresses listed as "Eye" Street. The designation, though, is meant to alleviate confusion between I Street and 1st Street.

K Street (NE, NW, SE, SW)
>   Originally the Old Post Road from Georgetown.

L Street (NE, NW, SE, SW)
M Place (SE, SW), Street (NE, NW, SE, SW)
N Place (SE), Street (NE, NW, SE, SW)
O Street (NW, SE, SW)
P Street (NW, SE, SW)
Q Lane (NW), Place (NW), Street (NE, NW, SE, SW)

1

Letter Streets

R Street (NE, NW, SE, SW)
S Street (NE, NW, SE, SW)
T Place (SE), Street (NE, NW, SE, SW)
U Place (SE), Street (NE, NW, SE, SW)
V Place (SE), Street (NE, NW, SE, SW)
W Place (NW, SE), Street (NE, NW, SE)

Named Streets

A

Adams Street (NW)
Formerly Albany Street, it was renamed in 1904 to honor the second and sixth presidents of the United States: father and son John (1735–1826) and John Quincy Adams (1767–1848).

Adams Place (NE)
See Adams Street.

Adams Mill Road (NW)
The mill was built by Benjamin Stoddert in the late 1700s. The property had three owners prior to being purchased by John Quincy Adams. At that time, the mill was given Adams's name. The buildings stood where Adams Mill Road bends into Rock Creek Park.

Akron Place (SE)
See City Streets.

Alabama Avenue (SE)
See State Streets.

Alaska Avenue (NW)
See State Streets.

Alden Place (NE)
James Alden (1810–1877) was a veteran of the Mexican and Indian Wars, but it was during his naval service in the Civil War that he was recognized for his bravery and leadership. As the commander of the steamer *South Carolina*, he served under Admiral David Farragut at the Battle of Mobile Bay (1864).

Ames Place (NE), Street (NE)
Adelbert Ames (1835–1933) was known as a formidable Union officer in the Civil War. He led infantry troops in the First Battle of Bull Run, the Antietam Campaign, the Battle of Chancellorsville and Gettysburg. His questionable politics after the war overshadowed his military record. Facing impeachment, Ames resigned as governor of Mississippi in 1876. Ames was the longest-surviving full-rank Civil War general.

Anacostia Avenue (NE), Freeway [I-295](SE), Road (SE)
The Anacostia section of the District of Columbia was originally named Twining and, later (1854), Uniontown. The latter name was found unsuitable as it could be confused with cities of the same name in Pennsylvania and Ohio. The name Anacostia to describe the area south of the Anacostia River was legalized in 1886. Anacostia was the first working class neighborhood in the District of Columbia.

Andrews Circle (SW)
General Frank M. Andrews (1884–1943) served in the Philippines and Hawaii during World War I. In 1917, he was transferred to a new aviation section of the Signal Corps. He was appointed commander of the first U.S. independent strategic unit, General Headquarters Air Force, in 1935. The success of the unit, which was instrumental in the development of the B-17 bomber, is credited to Andrews. At the start of World War II, Andrews served as commander in chief of the Caribbean theater. He was the first air officer to hold such a com-

mand. In February 1943, Andrews succeeded Dwight D. Eisenhower as commander of the U.S. forces in Europe. Andrews was killed three months later when his airplane crashed in Iceland. Andrews Air Force Base was named to honor the air force general.

Andrews Circle is located on Bolling Air Force Base.

Apple Road (NE)
See Flora and Fauna.

Arcadia Place (NW)
Arcadia is an area in southern Greece marked by barren, tortuous terrain. Through literature, Arcadia has gained a reputation as a paradise. The Greek poet Theocritus told stories of the fantastic people that lived there and Renaissance writer Sir Philip Sidney related romantic tales.

Arden Drive (NW)
William Shakespeare's *As You Like It* is, for the most part, set in Arden. Arden, which was once part of a great forest, is located in the western portion of Warwickshire, England. The area was often depicted in English literature as scenic and serene.

Argonne Place (NW)
This street is named to honor the battles fought at Argonne Forest. This heavily wooded ridge in northeastern France was the site of heavy fighting in both World War I and World War II.

Argyle Terrace
Argyle Terrace is named for a mill owned and operated by Thomas Blagden (see Blagden Street).

Arizona Avenue (NW)
See State Streets.

Arkansas Avenue (NW)
See State Streets.

Arnold Avenue (SW), Drive (NW)
Believe it or not, these streets are named for traitor Benedict Arnold (1741–1801). Arnold, serving as a general in the American Revolution, distinguished himself at Quebec and Saratoga. Passed over for promotion, Arnold set out to betray the American post at West Point. His plot, hatched with John Andre, was discovered. Arnold managed to escape and continue fighting the war for the British.

Asbury Place (NW)
Asbury Place is named for Francis Asbury (1745–1816), a Methodist bishop. Asbury arrived as a missionary from England and promoted the effective circuit rider system for spreading Methodism to the American frontier. Asbury's name can be found in place names around the country (for example, Asbury Park, New Jersey).

Ashby Street (NW)
Confederate Brigadier General Turner Ashby (1828–1862) was known for his personal bravery and his dedication to the Southern cause, but he was criticized as a commander. Ashby was killed leading a counterassault in the Shenandoah Valley. A soft-spoken man, after his death very few people could remember anything Ashby had said.

Presently, attempts are being made to create a monument to Turner Ashby in Harrisonburg, Virginia.

Aspen Street (NW)
See Flora and Fauna.

Astor Place (SE)
Astor Place is named for one of America's wealthiest families.

John Jacob Astor (1763–1848) monopolized the fur trade in the American Territories and dominated the market in China. After the Jay Treaty, Astor made arrangements to import his furs in Montreal. He expanded his operation to the West after the Louisiana Purchase. In 1808, he consolidated his holdings to form the American Fur Company. His plans in the Far West were foiled by the War of 1812. In 1813, he assumed the role of loan shark, loaning money to the government at astronomical interest rates. Astor retired in 1934, the richest man in America.

Astor's grandson, John Jacob Astor IV (1814–1912), fought in the Spanish-American War. He invented the bicycle brake. Astor built the Astoria half of the Waldorf-Astoria Hotel in New York City. He died when the *Titanic* sank on April 15, 1912.

A cousin, William Waldorf Astor (1848–1919), built the Waldorf half of the Waldorf-Astoria Hotel. He served as U.S. minister to Italy (1882–1885) and, unable to realize his political aspirations in the United States, he became a British subject in 1899.

Atlantic Street (SE)
Atlantic Street is named for the ocean of the same name.

Audubon Terrace (NW)
Audubon Terrace is named for ornithologist John James Audubon (1785–1851). Audubon came to the United States from present-day Haiti in 1803. Through extensive observation, he produced many paintings and drawings of birds, which were published in *The Birds of America*.

Austin Road (SE)
See City Streets.

Avon Lane, Place (NW)
No, Avon Lane and Avon Place are not named for the makeup salesperson who persistently rings doorbells. These streets, which run adjacent to each other, are

named for the river that flowed through the town of William Shakespeare's birth, Stratford-upon-Avon.

Azalea Road (NE)
See Flora and Fauna.

Bacon Drive (NW)
No, pigs have not run rampant in this area since Pierre L'Enfant's time. Bacon Drive, which runs behind the Vietnam Veteran's Memorial to the Lincoln Memorial, is named for Henry Bacon. Bacon designed the Lincoln Memorial as well as a memorial dedicated to those lost in the sinking of the *Titanic*.

Bancroft Place, Street (NW)
George Bancroft (1800–1891) served as Secretary of the Navy (1845–46) under James K. Polk. While in this office, Bancroft established the U.S. Naval Academy. Bancroft later served as minister to Britain (1846–49) and Prussia (1867–74). Bancroft gained recognition as an historian for his *History of the United States*, which, despite being pro-American, is considered a great piece of research.

Bangor Street (SE)
See City Streets.

Bank Street (NW)
Bank Street is named not for a local financial institution, but for its location . . . near the bank of the Potomac River.

Banneker Drive (NE)
Banneker Drive is named for Benjamin Banneker (1731–1806). Banneker's parents, slaves who had gained their freedom, sent him to school. Upon their death, they left him the few acres of land they owned. The land was situated next to property owned by the Ellicott family,

and Banneker, eager to gain knowledge about astronomy, spent much time devouring information found in books borrowed from George Ellicott. In 1791 George Washington requested Andrew Ellicott to survey the area that would be Washington. Ellicott asked Banneker to assist him with the initial calculations. A year later, Banneker began publishing an almanac that charted the positions of certain celestial bodies.

Barnes Street (NE)

Joseph K. Barnes (1817–1883), a member of the Army medical department, served with General Winfield Scott during the Mexican War. Barnes impressed Secretary of War Edwin M. Stanton, who appointed him acting surgeon general in 1863. One year later, the position officially became his. Barnes cared for Abraham Lincoln and Secretary of State William H. Seward after they were shot in 1865. He was one of the surgeons who fought unsuccessfully to save President James Garfield in 1881.

Barry Place (NW), Road (SE)

Commodore John Barry (c. 1745–1803) was the first naval officer commissioned under George Washington. In 1776, Barry was placed in command of the *Lexington* as a captain in the Continental navy. His capture of the British sloop *Edward* constituted the first American naval victory of the Revolution. When the ship he was commanding was caught in Philadelphia by the British blockade of the Delaware Bay, Barry fought with the army in Philadelphia, Trenton and Princeton. He later led a courageous run down the Delaware, which resulted in the capture of several British ships.

After the victory at Yorktown, Barry carried the Marquis de Lafayette home to France. In 1794, Washington asked Barry to train ensigns to form the core of the naval forces. He was awarded "Commission Number One" for his achievements. Barry served as commander of the naval station on Guadeloupe. At the time of his death in 1803, Barry, living in Philadelphia, was the senior naval officer.

Basin Drive (NW)
Basin Drive is best known in the spring for the cherry trees that line the street as it travels around the southern side of the Tidal Basin.

Bataan Street (NW)
Located on the western side of Scott Circle, Bataan Street is named to honor U.S. and Filipino troops captured on the Bataan peninsula by the Japanese during World War II. These troops were led on a brutal "death march" to a prison camp.

This street and Corregidor Street were nameless until Philippine ambassador Carlos P. Romula requested they be named in 1961. Residents whose houses faced these nameless streets were thrilled because the naming resolved the problem of proper addresses.

Bates Road (NE), Street (NW)
Edward Bates (1793–1869) was an anti-secessionist who decided to support Abraham Lincoln after an unsuccessful run for the presidency against him. Lincoln appointed Bates attorney general, but Bates disagreed publicly with Lincoln on many issues. His controversial stance cost him support within the government; he resigned his post in 1864.

Battery Place (NW)
Battery Place runs adjacent to Battery Kemble Park and is named for this Civil War fortification.

Beach Drive (NW)
Beach Drive runs through Rock Creek Park from Connecticut Avenue to the Maryland border and beyond. The street is named for Lansing H. Beach, one of the first men responsible for the creation of Rock Creek Park.

Beech Street (NW)
See Flora and Fauna.

Beecher Street (NW)
Beecher Street is named for Henry Ward Beecher (1813–1887). An outspoken and often theatrical minister, Beecher preached against slavery. Unlike many abolitionists of the time, Beecher believed abolishing slavery would be unconstitutional; he felt that the growth of slavery should be halted and its evils made known until it ceased to exist. Beecher, who often shared the spotlight with his sister Harriet Beecher Stowe, was known for lending his support to unpopular causes.

Beechwood Road (NE)
See Flora and Fauna.

Bellair Place (NE)
It is highly likely that Bellair Place, French for "good air," takes its name from the Maryland city of the same name.

Bellevue Street (SE), Terrace (NW)
Bellview Drive (SW)
The above two streets (Bellevue and Bellview) derive their names from the French for "good view." Since many of the scenic outlooks in Washington have been obstructed by architectural progress, it is sometimes difficult to imagine what the namers saw. Bellview Drive has a view of the Potomac River from its location on Bolling Air Force Base. Unfortunately, its close proximity to the Blue Plains Sewage Disposal Plant makes the scent less than desirable.

Belmont Road (NW), Street (NW)
Both of these streets are built on hilly terrain. The name comes from the French, meaning "good mountain."

Belt Lane (NW), Road (NW)
Colonel Joseph Belt, a member of the Maryland House of Burgesses, led the Prince George's County militia

during the French and Indian War. Although not known for any spectacular performance in the military or political arenas, Belt is known for a different reason: The 560 acres he owned became today's Chevy Chase. Belt Road, which traversed his property, has been cut into several sections as it takes a diagonal course across the Friendship Heights area of Washington. The farmhouse that Belt built on his property still remains as part of the Chevy Chase Country Club. (See Chevy Chase Circle.)

Bending Lane (NW)
Although it may have followed a twisting course at one time, Bending Lane's one-block journey is very straight today.

Benning Road (NE, SE)
Nicknamed "Old Rock," Henry L. Benning (1814–1875) repeatedly distinguished himself in battle. Benning served as an associate justice of the Georgia Supreme Court until the start of the Civil War. Benning fought fearlessly in battle and led troops to Confederate victories at Sharpsburg and Fredericksburg. Fort Benning, Georgia, occupied by the U.S. Army, is named in his honor.

Benton Place (NW), Street (NW)
This street is named for the Civil War gunboat USS *Benton*. The boat was constructed by James B. Eads and saw much action during the war. The boat was initially commanded by Andrew H. Foote. While under David Porter's command, the ship took part in battles on the Yazoo River, at Vicksburg and during the Red River Campaign. The ship was scrapped at the end of the war.

Berkley Terrace (NW)
See City Streets.

Berry Road (NE)
See Flora and Fauna.

Bingham Drive (NW)
After Washington was built and had begun to function as the seat of the federal government, interest in Pierre L'Enfant's original design waned. Colonel Theodore A. Bingham, Chief of Public Buildings and Grounds realized the need for more organized development. At the turn of the twentieth century, he began actively campaigning to redevelop the Mall according to L'Enfant's original plans.

Birch Drive (NW), Street (NW, SE)
See Flora and Fauna.

Birney Place (SE)
Abolitionist James Gillespie Birney (1792–1857) was never in agreement with slavery, but he never opposed it either. He became an agent for the American Colonization Society, and by 1834 he had changed his original ideas about slavery. In 1835, he organized the Kentucky Anti-Slavery Society after freeing his own inherited slaves. He published the *Philanthropist* from 1836–1837, despite opposition. Birney became executive secretary of the American Anti-Slavery Society in 1837, but his ideas for political action differed from those of William Lloyd Garrison and he struck out on his own aided by the creation of an anti-slavery party (see Garrison Street).

Birney was nominated by the Liberty Party for the presidency twice (1840, 1844). In 1840, Birney only managed to attract a small amount of votes. In 1844, though, he managed to swing support away from Henry Clay, allowing James K. Polk to win the state of New York.

Bladensburg Road (NE)
Bladensburg Road was the main road north to Bladensburg, Maryland, and points beyond. Bladensburg was named for Thomas Bladen, Maryland's governor from 1742 to 1747.

Blagden Avenue, Terrace (NW)
Thomas Blagden operated a flour mill and a bone mill. The mill was sometimes called Argyle Mills.

Blaine Street (NE)
James G. Blaine (1830–1893) was a spirited politician, serving as Maine's political "boss" for 32 years. In 1854, he edited the *Kennebec Journal,* making it a Republican newspaper and establishing the party in Maine. He served as a congressman (1862–1868), Speaker of the House (1868–1876) and Secretary of State (1888–1892) under William Henry Harrison, James Garfield and Benjamin Harrison. While holding the latter office, Blaine chaired the first Pan-American Conference in 1889. He was a strong Lincoln supporter, urging leniency toward former Confederate states and advocating giving blacks the right to vote. Blaine was eager to make his home in the White House, but he ran three unsuccessful campaigns for president. Running against Grover Cleveland in 1884, he was tagged as "Blaine, the Liar from the State of Maine." Blaine lived out his years in a Dupont Circle mansion built for him in 1881. Blaine Mansion still stands at 2000 Massachusetts Avenue.

Blair Road (NE, NW), Portal (NW)
Andrew Jackson requested the presence of Francis Preston Blair (1791–1876) in Washington. Blair's duty was to found *The Globe,* the newspaper that would become the voice of the Jackson administration. Blair served in Jackson's "Kitchen Cabinet." He was considered one of the most influential political journalists in the country. He also published the *Congressional Globe,* predecessor to the *Congressional Record.*

Blair supported Western expansion, without the similar expansion of slavery. In 1848, he split with the Democrats and lent his support to Martin Van Buren and the Free Soil Party. As the slavery issue developed, Blair felt that the Democrats had shifted from their original purpose. To remedy this and regain the original party beliefs, Blair helped create the Republican party in 1856. He actively supported Abraham Lincoln in 1860 and, for his support, earned a position as his adviser. When the Democrats he opposed became powerful, Blair rejoined that party.

Blanchard Drive (SW)
Albert G. Blanchard (1810–1891) graduated from West Point and served in the army until 1840. He returned to military service for the Mexican War and the Civil War, which he fought under the Confederate flag. Known as an ineffective commander and branded incompetent by his superiors, Blanchard was given secondary assignments in Virginia.

Blue Plains Drive (SW)
In 1663, a land patent was issued to George Thompson for Beau Plaine. In 1714, the Prince George's County plantation, which was located near today's Blue Plains Drive, was named Bew Playnes. Bew (Beau) was corrupted to Bleu and then Blue over two centuries and Plaine took on a more modern spelling. Today, the land is home to the Blue Plains Sewage Disposal Plant.

Bowen Road (SE)
John Stevens Bowen (1830–1863) resigned from the army in 1856 to become an architect in St. Louis, but reconsidered when the Civil War broke out. Wounded at Shiloh, he returned to fight in all the battles leading up to the Siege of Vicksburg. Bowen served as the communication link between Lt. Gen. John C. Pemberton and Maj. Gen. Ulysses S. Grant, convincing Pemberton to surrender to the Union troops to end the siege. Bowen died 10 days after the surrender.

Brandywine Place (SE), Street (NW, SE)
This street is named for the Battle of Brandywine, which took place on September 11, 1777. The battle, fought between Sir William Howe and George Washington, took place along the banks of the Brandywine Creek in

Pennsylvania. Washington was forced to retreat, allowing the British to advance and take Philadelphia.

Brentwood Parkway (NE), Road (NE)

Thomas Jefferson appointed Robert Brent to be Washington's first mayor, a position he held from 1802 to 1812. Brent established many new programs. Much to the delight of the residents, he imposed taxes. On the brighter side, he created the first fire department and opened the city's first two public schools.

Brent's country home, Brentwood, was designed by Benjamin Latrobe and lends its name to both streets, which are continuations of 6th, 9th and 13th Streets. The house, built in 1816 at what is now 5th Street and Florida Avenue, was razed in 1917.

Broad Branch Road (NW)

Broad Branch Road takes its name from a tributary of Rock Creek.

Bryant Street (NE, NW)

Formerly Baltimore Street, the name was changed in 1905 amid much resident furor. Angry letters were received by the commissioners requesting that the name remain Baltimore because it had more class than Bryant.

Buchanan Street (NE, NW)

Once Vallejo Street, it was renamed in 1905 for the 15th president of the United States, James Buchanan (1791–1868). Buchanan, the only bachelor president, saw slavery as morally wrong, but constitutional. He attempted to keep the "sacred balance" between pro- and anti-slavery factions. After his re-election defeat, several states threatened secession; he promised no violence would occur during negotiations. However, as a precautionary measure, U.S. troops were sent to Fort Sumter; the Civil War began shortly after his departure from office.

Buckeye Drive (SW)
Buckeye Drive joins the eastern and western branches of Ohio Drive, which circles East Potomac Park. Fittingly, the buckeye is the state tree of Ohio. (See Flora and Fauna.)

Bunker Hill Road (NE)
This road is named for the Revolutionary War Battle of Bunker Hill. The battle took place on Breed's Hill in Charlestown, Massachusetts, on June 17, 1775. The American militia stymied British General William Howe's advance . . . until they ran out of powder. This British victory failed to break the American hold on Boston.

Burbank Street (SE)
This street is named for the creator of the Shasta daisy and the Burbank potato, Luther Burbank (1849–1926). A plant breeder, Burbank developed many varieties of fruits, vegetables and flowers.

Burlington Place (NW)
See City Streets.

Burns Place, Road, Street (SE)
"The obstinate Mr. Burnes," as George Washington referred to him, David Burnes (d. 1799)—or Burns, as it was often spelled—was one of the primary land owners of the future Washington. A successful farmer and county official, Burnes saw the economic benefits he could reap from the building of the capital city and was one of the first to take advantage of the opportunity. The Burnes-owned property, bordered by today's 3rd Street, H Street, 18th Street and Constitution Avenue, was earmarked for the President's House. Burnes refused, whether due to slick bargaining techniques or plain orneriness, to transfer his property to the government. He finally conceded, with the stipulation that his house remain untouched by the construction of the new city. He received the highest price per acre paid by the gov-

ernment. Burnes complained that his crops were damaged by Pennsylvania Avenue being cut through his fields and by the construction of the White House.

Upon his death, Burnes's daughter Marcia inherited his land, making her "the richest girl in America." As she was still unmarried, many men tried to win her hand. In 1802, she married Congressman John Peter Van Ness (see Van Ness Street). Burnes's house stood until 1894.

Butler Street (SE)

General Benjamin F. Butler (1818–1893), commander of Fort Monroe, Virginia, declared in May 1861 that any escaped slaves who managed to reach his lines would be considered "contraband of war." He stated that under no circumstances would these individuals be returned to their masters. Eleven months later, on April 16, 1862, slavery was outlawed in the District of Columbia.

While commanding forces in New Orleans, Butler claimed to have single-handedly halted crime in the city. A band of men, masquerading as Union troops, were found to be ransacking houses. Butler, discovering that the band had gathered in a local pub, ordered all the patrons of the tavern brought before him. He recognized two of the men and, after receiving confessions, had all but one of the conspirators hanged. He stated that after the men had been brought to justice, there were no more burglaries in the city of New Orleans.

Back in Washington after the Civil War, Butler led the movement to impeach Andrew Johnson in 1868. In 1884, he made his own run for the presidency as the Greenback Party candidate.

Butternut Street (NW)

Butternut was a nickname for Confederate soldiers because at one time their uniforms were yellowish-brown. (See Flora and Fauna.)

California Street (NW)

During the reassignment of street names in 1905, California Street was renamed T Street. In 1906, bills were passed in Congress to retain the original name between Columbia Road and Massachusetts Avenue, to properly honor the state name in the District of Columbia. California is one of two states not represented by an avenue.

Calvert Street (NW)
Named for the Calvert family, the street honors Charles, George, Cecilius and Leonard Calvert.

George Calvert, first Baron Baltimore (1580–1632) was granted territory by James I in 1632. Calvert's land is now Maryland and the District of Columbia. Calvert had a dream of creating a new colony, but he died before the charter could be approved. The charter was officially transferred to his son Cecilius (1605–1675). Cecilius, the second Baron Baltimore, never visited the colony.

Charles Calvert, third Baron Baltimore (1637–1715), was appointed by his father, Cecilius, governor of the colony of Maryland in 1661.

The son of George Calvert, Leonard (1606–1647) was sent by Cecilius to America with two shiploads of colonists. Once on land, Leonard established St. Marys (St. Marys City, Maryland). Calvert, whose government started well, had a dispute with some local residents. William Claiborne had a trading post on Kent Island; Calvert claimed the island in 1638 and forced Claiborne out. In 1644, Claiborne collected a force of Virginians and captured St. Marys. Calvert spent two years in exile in Virginia; he retook the city upon his return.

Camden Street (SE)
See City Streets.

Canal Road
Canal Road travels along a portion of the 185-mile Chesapeake and Ohio Canal, from George Washington Memorial Parkway near the Maryland border to M Street in the heart of Georgetown.

Ground was broken for the digging of the canal on July 4, 1828. The port at Georgetown was beginning to suffer; construction of the canal was seen as the port's last chance for survival. John Quincy Adams said the canal would be "a conquest over physical nature such as has never been achieved by man." Two years after work began, the canal stretched 184 miles to Cumberland, Maryland, with barges transporting coal, wood and grain.

The canal was destined to fail, though. On the same day in 1828, Charles Carroll of Carrollton broke ground for another, more modern method of transportation: the railroad. The Baltimore and Ohio, America's first railroad, provided a much faster means of transporting materials to destinations across the United States. The traffic on the canal diminished until it became the recreation area that it is today.

Canal Street (NW, SE, SW)

Canal Street, all that remains of the polluted course of the failed Washington Canal, runs a choppy V-shaped course through Southeast and Southwest Washington. On the western side, it begins at the banks of the Anacostia River and travels through Fort Lesley J. McNair until it terminates abruptly at N Street, SW. The eastern aspect proceeds north from the entrance of the Navy Yard to I Street, where it jogs westward and loops toward I-395. The street used to take a short diagonal excursion from South Capitol Street to Independence Avenue, but in 1989 this section was renamed Washington Avenue.

The canal, a combination of the existing Tiber and St. James Creeks, was meant to be a commercial as well as a recreational waterway. Pierre L'Enfant envisioned the canal as one of the focal points of the federal city. The Washington Canal was built by Benjamin Latrobe. Beginning at two separate points on the Anacostia River, the canal travelled north until the two branches met south of the Capitol grounds. The canal continued north and west to bend around the present day Mall. (See Constitution Avenue.)

Carroll Street (NW)

Carroll Street is named for the Carroll family, who not only owned much property in the new capital city, but were instrumental in guaranteeing its success.

Daniel Carroll of Carrollton was chosen by George Washington to be a commissioner overseeing the new city.

Daniel Carroll of Duddington, nephew of the previous Carroll, was president of the Bank of Washington and the Columbian Turnpike Company. Carroll built the first tavern in the District of Columbia, located on what are now the Capitol grounds. The tavern was burned by the British in 1814. Carroll also constructed a three-building hotel for Congress. The property, Carroll Row, stood until the 1880s. The Library of Congress now occupies the land. Carroll owned the property and house on New Jersey Avenue that served as a catalyst for Pierre L'Enfant's downfall, although both Daniels protested his dismissal from the commission. (See Duddington Place.)

Charles Carroll III of Carrollton was a Maryland planter, businessman and politician. Carroll was the longest surviving signer of the Declaration of Independence. He was also the only Roman Catholic to sign the document.

John Carroll (1735-1818) was the first American Roman Catholic archbishop. He founded Georgetown University in 1789.

Carrollburg Place (SW)

Charles Carroll (1737-1832) had a dream to build a port city, which he would name Carrollsburg, at the mouth of the Anacostia River. Carroll acquired property in 1759 and built his house, one of the first built in the Anacostia neighborhood, near today's Navy Yard (at South Capital and Q Streets and Potomac Avenue). Although Carroll's dream was never realized, he instituted the Carrollsburg name in the area and Carrollburg Place, four blocks north of his house, bears evidence of this.

Cathedral Avenue (NW)

Cathedral Avenue emerges from both sides of the National Cathedral. In 1813, Joseph Nourse purchased 82 acres of Pretty Prospect on Mt. Alban. (See Prospect Street.) Nourse, who served as Registrar of the Treasury (1789-1829), took the name from Mount Saint Alban in his birthplace of Herfordshire, England. In his will,

Cathedral Avenue (NW)

Nourse expressed his desire that a religious institution be placed on his land. The cathedral was only a hope until Henry Y. Satterlee, the first Episcopal bishop of Washington, made plans for the cathedral. At the turn of the twentieth century, work began on the structure. The groundbreaking was conducted by Theodore Roosevelt in 1907.

Planned, but never completed, the cathedral's second tower was built in 1990. Woodrow Wilson, the only U.S. president buried in Washington, is interred in the cathedral. Others buried at the cathedral include Bishop Clagett, Bishop Satterlee, Admiral George Dewey, Helen Keller and Anne Sullivan.

Caton Place (NW)

Caton Place is named for Richard Caton, the husband of Charles Carroll's daughter. Caton was a failure in business, a career that ended in bankruptcy. Charles Carroll gave the couple land, now known as Catonsville, Maryland, which firmly planted the Caton name in history.

Cedar Drive (SE), Street (NW, SE)

See Flora and Fauna.

Chain Bridge Road (NW)

Ironically, Chain Bridge Road does not lead to Chain Bridge. The street begins at Nebraska Avenue and dead-ends just short of Canal Road; it continues off the bridge into Virginia. At one time, Chain Bridge Road provided a continuous route to the bridge, but with the creation of newer roads and housing and recreational developments, the direct course was obliterated.

Champlain Street (NW)

Champlain Street is named for the fourth largest freshwater lake in the United States, Lake Champlain. The lake forms part of the border between New York and Vermont.

The lake bears the name of French explorer Samuel de Champlain (1567–1635). Champlain founded Port Royal, Nova Scotia, and settled Quebec, Ontario.

Channing Street (NE)
Formerly Cincinnati Street, the street name was changed in 1905 to honor William Ellery Channing (1780–1842). A minister, Channing advocated humanitarianism. Including writings on almost every political subject, Channing's work illustrated progressive ideas about slavery, war and education.

Chanute Place (SW)
Chanute Place is named for engineer Octave Chanute (1832–1910). After gaining a reputation as a builder of railroads and railroad bridges, Chanute, in 1867–8, constructed the first bridge over the Missouri River. The bridge was erected in Kansas City.

Chanute was also interested in aviation. In 1894, he wrote *Progress in Flying Machines.* He made over two thousand glider flights at his camp on Lake Michigan near Chicago. Through his research, he improved glider designs. Orville and Wilbur Wright's 1903 flight at Kitty Hawk was aided by Chanute's data and designs.

Chapin Street (NW)
Resident August Peterson complained in 1901 about the proposed change of Chapin Street to Zane Street. Peterson stated that he had looked Zane up in the local directory and discovered only one. He claimed he had no desire to live on a street named for a Treasury employee living in Anacostia. Although his facts were flawed, the street name was never changed.

Charleston Terrace (NW)
See City Streets.

Chase Circle (NW)
Chase Circle is named in honor of Salmon P. Chase (1741–1811). Chase served as a U.S. senator (1849–55,

Chase Circle (NW)

1861) and as governor of Ohio (1855–59). Chase, as Secretary of the Treasury under Abraham Lincoln, created a national bank system. As chief justice of the Supreme Court (1864–73), he presided over the impeachment proceedings against Andrew Johnson.

Cherry Road (NE), Street (SE)
See Flora and Fauna.

Chesapeake Street (SE, NW)
This street is named for the Chesapeake Bay. The waters of the Potomac dump into the bay on the eastern coast of Maryland.

Chestnut Street (NE, NW)
See Flora and Fauna.

Chevy Chase Circle, Parkway (NW)
No, Chevy Chase was not named for the comedian. The Chevy Chase area, both in Maryland and in Washington, is named for a tract of land owned by Col. Joseph Belt. He named his property Chevy Chase. The Chevy Chase Land Company, which was formed in 1890, began the first substantial development of the area. Sixteen houses in the area were built prior to 1909. The circle, only one-third of which is inside the District boundary, and the parkway bearing the Chevy Chase name took it from the area. (See Belt Road.)

Chicago Street (SE)
See City Streets.

Church Street (NW)
Captain B.S. Church of the Engineer Corps led the first Union regiment into Confederate territory after the start of the Civil War. His troops crossed the Long Bridge into

Virginia on May 24, 1861. Church Street was later home to the Foundry Methodist Church, founded by Henry Foxall (See Foxhall Street).

City Streets
Akron (Ohio) Place
Bangor (Maine) Street
Berkley (California) Terrace
Bladensburg (Maryland) Road
Cambridge (Maryland) Place
Camden (New Jersey) Street
Charleston (South Carolina) Terrace
Chicago (Illinois) Street
Cleveland (Ohio) Avenue
Cumberland (Maryland) Street
Denver (Colorado) Street
Elmira (New York) Street
Flint (Michigan) Place
Frankfort (Kentucky) Street
Frederick (Maryland) Place
Gainesville (Virginia) Street
Galveston (Texas) Street
Greenwich (Connecticut) Parkway
Hartford (Connecticut) Street
Joliet (Illinois) Street
Kanawha (West Virginia) Street
Kensington (Maryland) Place
Laurel (Maryland) Street
Lexington (Virginia) Place
Luray (Virginia) Place
Newark (New Jersey) Street
Newport (Rhode Island) Place
(New) Orleans (Louisiana) Place
Portland (Oregon) Street
Princeton (New Jersey) Place
Providence (Rhode Island) Street
Quebec (Canada) Street
Raleigh (North Carolina) Street
St. Louis (Missouri) Street
Salem (Virginia) Lane
Saratoga (New York) Avenue
Savannah (Georgia) Street
Sudbury (Massachusetts) Road

Trenton (New Jersey) Street
Williamsburg (Virginia) Lane
Wilmington (Delaware) Place
Xenia (Ohio) Street
Yorktown (New York) Road
Yuma (Arizona) Street

Clagett Place (NE), Street (SW)

Capt. Thomas Clagett settled land in 1682 in what is now northeast Washington and named it Clagett's Delight.

John Clagett served as a vestryman in the Rock Creek Parish. "Out of sentiments of tenderness and regard for the rising generation" he donated parish land for the construction of a public school.

Clagett Street is named for the family, whose name is thought to be derived from Thomas' birthplace, Claygate, England.

Clay Place (NE), Street (NE), Terrace (NE)

Called the Great Pacificator and the Great Compromiser, Henry Clay (1777–1852) was a tireless politician. He served as Secretary of State (1825–29) to John Quincy Adams as well as several terms as Congressman from Kentucky (1806–7, 1810–14, 1815–21, 1823–25, 1831–42, 1849–52). He helped initiate the War of 1812 and pushed the Missouri Compromise through the House of Representatives. Clay's Compromise outlawed slave trade in Washington, D.C.

Clermont Drive (NE)

In 1807, Robert Fulton launched his steamship *Clermont* on the Hudson River. The boat's commercial success led him to be considered the inventor of the steam-powered ship. Clermont Drive is named to honor Fulton's invention (See Fulton Street).

Cleveland Avenue (NW)

Cleveland Avenue is named for Grover Cleveland (1837–1908), the 22nd (1885–1889) and 24th (1893–1897) President of the United States. In the 1884 election, Cleveland defeated the Tammany Hall representative, James G. Blaine. In 1888, he lost to Benjamin Harrison, the same candidate he defeated in 1892. Cleveland, noted for his honesty, faced financial problems during his administration. He is also noted for breaking the Pullman strike of 1894.

Cliffbourne Place (NW)
Prior to the massive renaming of streets in 1905, this half-block street was considered part of 19th Street. The residents petitioned the commissioners, requesting that the street be named Cliffbourne or Cliffburn Place. Debates raged between residents as to which spelling should be placed on the records. Some stated that Cliffburn was the correct spelling of an old Washington family name; others claimed that Cliffbourne was more convenient due to names already posted on the corner apartment building and pharmacy. In a letter to the commissioners dated July 25, 1905, Robert Lincoln O'Brien claimed, "The object of a name is not to perpetuate the spelling of a family [name] comparatively unknown to the present generation in our neighborhood, but to make it easy for people to find where we live." The Cliffbourne supporters claimed victory in October 1905.

Clinton Street (NE)
George Clinton (1739–1812) was Vice President of the United States (1805–12) under James Madison until his death. A strong advocate of state sovereignty and opposer to a federal constitution, he served as the first governor of New York under the new state constitution (1777–95).

College Street (NW)
College Street, aptly named for its location, traverses the Howard University campus.

Colorado Avenue (NW)
See State Streets.

Columbia Road (NW)

Named for Columbia University, Columbia Road is one of six streets in the original Columbia Heights area to be named for Ivy League colleges. The street, composed of old farm roads, could not conform to the Washington street plan because, the commissioners stated, "it turns south and crosses a number of east/west streets."

Commodore Joshua Barney Drive (NE)

Baltimore native Joshua Barney (1759–1818) first went to sea at age 11. At 15, he successfully completed a voyage when the commander died while the ship was in the middle of the Atlantic Ocean.

In 1775, Barney joined the colonial navy. During the Revolutionary War, he served on many ships and performed some incredible nautical maneuvers in the defense of America. In 1794, disgusted with the rank given him, Barney resigned as captain of a ship being sent to battle the Barbary pirates. Two years later he accepted a commission in the French navy, where he served for six years.

During the War of 1812, Barney commanded privately owned ships against British shipping. In 1814, his presence was requested in Washington to aid in the defense of the city. His armed barges along the Chesapeake Bay halted the British advance for a time, but the royal force eventually eluded them. Barney marched his troops to Bladensburg, Maryland, mounted ship's guns on carriages and placed them at the center of the American position. The Americans were defeated, but Barney's troops provided some resistance. Barney was wounded in the battle.

Congress Court (NW), Place (SE), Street (NE, SE)

Believe it or not, none of the above streets is near the Capitol. The closest is Congress Street, NE (it runs for one block along the railroad tracks leading out of Union Station). Congress Street, SE, and Congress Place are located in Anacostia and Congress Court can be found in Georgetown. This illustrates the impact of the political body on the entire city.

Conifer Road (NE)
See Flora and Fauna.

*Connecticut Avenue (NW)
See State Streets.

Constitution Avenue (NE, NW)
Constitution Avenue, with its path past many of Washington's memorials and museums, is considered one of Washington's major parade routes. Unlike the funeral processions and inaugural and victory parades that progress down Pennsylvania Avenue, the parades that travel Constitution Avenue are less ceremonial, for example, the Cherry Blossom Parade and the Gross National Parade. Franklin D. Roosevelt was the only president to have his inaugural parades down this street. The American Institute of Pharmacy is the only private building situated on Constitution Avenue.

One of L'Enfant's original plans called for a canal to traverse the city from the Capitol to the Anacostia River (Eastern Branch). The canal was to be a point of beauty as well as a recreation area on the Mall. In 1810, Benjamin Latrobe began constructing the Washington Canal, a combination of the Tiber and St. James Creeks. During its construction, local merchants were praising the future benefits of the Canal, but by 1826, the canal had become too difficult to navigate due to its unsuitable width and depth (it had to be dredged constantly) for the boats of the day. As traffic on the Canal declined, it became an unsightly health hazard and by the early 1870s it was filled. The section that ran from the Capitol around the Mall was named B Street. In 1931, it was renamed Constitution Avenue after the names Memorial and Jefferson were rejected. A lock house at Constitution Avenue and 17th Street is the only physical evidence that remains of the failed Washington Canal (See Canal Street).

Corcoran Street (NE, NW)
William Wilson Corcoran (1799–1888), founder of the Corcoran Gallery of Art, was known as "Washington's First Philanthropist." Corcoran, who always sported a

red rose in his lapel, inherited a fortune from his father, a two-time mayor of Georgetown. He increased his wealth as a founder of the Corcoran and Riggs bank in 1840 (now Riggs Bank). He was generous with his wealth, donating money to area universities and assisting the poor.

Corcoran was interested in making Washington a better place to live. After William Henry Harrison's death in 1841, he helped the government get much-needed loans. He secured loans for the government again during the Mexican War. He also helped secure U.S. credit abroad, borrowing money from London banks. Corcoran established Oak Hill Cemetery in honor of his wife. Corcoran felt that Pierre L'Enfant's contribution to Washington's creation was substantial and he attempted to have his remains interred in the District. His attempts were denied by Congress.

The first home of Corcoran's art collection, located at 17th Street and Pennsylvania Avenue, was designed by James Renwick. The building was completed in the early 1860s at a time when the Civil War had placed the government in desperate need of office space. Congress, knowing of Corcoran's southern sympathies, commandeered the building for General Montgomery Meigs's headquarters before any exhibits could be displayed for public viewing.

Corcoran travelled to Europe during the Civil War. He returned to Washington in 1866 to charges of tax evasion. The case, brought by Edwin Stanton, was dropped due to lack of evidence. The government graciously returned the art gallery building after the War ended (in 1870, to be exact), but not in the same shape they had received it. In 1871, the gallery opened. The ball held at the gallery not only celebrated the long-awaited exhibition, but also raised money for the completion of the Washington Monument. (See Harewood Road.)

Corcoran was a popular figure in Washington. The citizens of the capital mourned his death: "He was so much to Washington, and Washington was so much to him that the man and the city seemed indissolubly associated . . . no other name, except that which the Capital bears, no other memory except that of the Father of our Republic, are so dear to the hearts of the people of this city."

Corregidor Street

A well-fortified island off the Bataan peninsula in Manila Bay, Corregidor was the site of five months of Japanese bombardment during World War II. Corregidor Street, located on the eastern side of Scott Circle, honors the ten thousand U.S. and Filipino troops forced to surrender to the Japanese in 1942 and the U.S. recapture of the island in 1945.

Philippine ambassador Carlos P. Romula requested that two nameless streets be renamed Corregidor and Bataan Streets in 1961. Many residents of these formerly nameless streets were thrilled because the naming gave them addresses.

Crabtree Road (NE)

See Flora and Fauna.

Crescent Place (NW)

Crescent Place is a descriptive name for this block-long, curved street, which runs north from Meridian Hill Park.

Crittenden Street (NE, NW)

Crittenden Street is named for a U.S. senator from Kentucky, John J. Crittenden (1787–1863). The senator supported the Crittenden Compromise (1860), an amendment to the Missouri Compromise. The proposal extended the Missouri Compromise border to California, dividing free and slave states. The Crittenden Compromise, an attempt to prevent the Civil War, was defeated by Congress.

Cromwell Terrace (NE)

Cromwell Terrace is named to honor Oliver Cromwell (1599–1658). In 1653, after distinguishing himself in military service, Cromwell was appointed Lord Protector of England. In 1657, he was offered the throne, but he refused. Considered a military genius, Cromwell exhibited tendencies toward cruelty and violence.

Cumberland Street (NW)
See City Streets.

Cushing Place (NW, SE)
Caleb Cushing served as attorney general (1853–57) under Franklin Pierce.

Dahlia Street (NW)
See Flora and Fauna.

Danbury Street (SE)
See City Streets.

Davenport Street (NW)
Davenport Street is named for inventor Thomas Davenport (1802–1851). The son of a blacksmith, Davenport decided to go into the business. Although he was quite successful, Davenport's attention was drawn by an electromagnet. He experimented with its force and determined that the magnet could be used to create power. In 1834, he created an electric motor, the predecessor of today's motors. Davenport found that the motor and some of his other inventions (including a working model of an electric trolley) attracted little attention. He published journals, printing them on his electric press, but he was forced to cease his promotional activities due to lack of funds. His final creation was an electric player piano.

Deane Avenue (NE)
Deane Avenue is named for Silas Deane (1737–1789). Deane was a leader in the Revolutionary movement in Connecticut. He served in the first Continental Congress in 1774 and again in 1775. Deane sat on the committees that organized the navy. The following year, Deane was not elected to the Congress, so he was sent to France with two duties. The first was an attempt to receive military assistance; the second involved selling

American produce in return for much-needed supplies. Deane met with Edward Bancroft, an old friend, in Paris as he had been instructed. With his sights set on profiteering, Deane informed Bancroft of American and French secrets. Unfortunately, Bancroft was working for the American and British governments. Another assignment given to Deane by the Continental Congress was to find four engineers for the army. Deane did not know what qualifications were necessary for the selection. He made contracts with most of the foreign officers who expressed an interest in the cause of the Continental army. In 1776, Congress appointed two other men, Benjamin Franklin and Arthur Lee, to assist Deane with his diplomatic mission. Congress ratified the French Alliance in May 1778.

Deane came back to America, but after two years he returned to Europe. In 1781, he wrote to friends and stated that his support for independence was waning. The British, aided by Bancroft, published the contents of the letters, stating that they had been intercepted. Deane was accused of treason. He spent the last eight years of his life in exile abroad. In 1842, Congress gave Deane's heirs $37,000 as partial restitution for his war expenses. This gesture helped to improve the diplomat's reputation.

Decatur Place, Street (NE, NW)

American naval hero Commodore Stephen Decatur (1779–1820) was known for his unyielding bravery and patriotism during the War of 1812. He used the prize money he received from his victories over the Barbary pirates and British in the war to build the first private house on President's Square (see Lafayette Square). Decatur was killed in 1820 in a duel at Bladensburg with Commodore James Barron, who blamed him for damaging his career and reputation. On March 22, the *National Intelligencer* published a eulogy that proclaimed, "Mourn Columbia! for one of thy brightest stars is set!"

Defense Boulevard (SW)

Defense Boulevard is located on the grounds of Bolling Air Force Base.

Delafield Place (NE, NW), Street (NE)

Delafield Street is named for engineer Richard A. Delafield, who served during the Civil War with the rank of brigadier general.

Delaware Avenue (NE, SW)

See State Streets.

Dent Place (NW)

As Ulysses S. Grant's brother-in-law, Frederick T. Dent (1820–1892) received some special treatment. When Grant assumed command of the Union forces, he brought Dent onto his staff to command the volunteer troops. Dent served a short term as the military governor of Richmond, then returned to Washington to command its defenses and serve as President Grant's military secretary.

During the renaming of streets in 1905, the Georgetown Citizen's Association was given great freedom. Originally named S Street, the commissioners recommended changing the name to Cambridge Place. The residents did not like the name and suggested several, including Irving Place, that were denied due to duplication. In June 1905, the residents decided upon the acceptable name of Dent Place.

Denver Street (SE)

See City Streets.

Dexter Street (NW)

Dexter Street is named for Samuel Dexter. Dexter served under John Adams and Thomas Jefferson as secretary of the treasury and under Adams as secretary of war.

Dix Street (NE)

Dix Street is named for John Adams Dix (1798–1879), the soldier and statesman for whom Fort Dix, New Jersey, is named. After seeing some action in the War of 1812, Dix became active in New York politics. He became state adjutant general in 1830, followed by a term as Secretary of State of New York (1833–1839). He entered federal politics in 1845 when he was chosen to fill a vacant Senate seat. A prominent anti-slavery Democrat, Dix lent his support to Martin Van Buren and the Free Soil Party in 1848. That year, he also ran unsuccessfully for the governorship of New York.

When his Senate term expired in 1849, Dix left politics and became a force in the railroad business, serving as president of the Union Pacific and the Erie Railroads.

Although he received several nominations for appointed office, Southern Democrats halted the action. In 1860, after a scandal, Dix was appointed Postmaster of New York City by James Buchanan. A smart businessman, Dix served two months as Secretary of the Treasury and instilled confidence in bankers. As the Civil War hostilities were rising in the south, Dix delivered this message to the Treasury office in New Orleans: "If anyone attempts to haul down the American flag, shoot him on the spot." Due to his advancing age, Dix served mostly as military administrator during the Civil War, although he did lead a few brigades of volunteers into minor combat situations.

After the war, Dix served three years as minister to France. In 1872, although still a member of the Democratic party, Dix was nominated by the Republicans for governor of New York. He won by a landslide and served one term.

Dogwood Street (NW)
See Flora and Fauna.

Douglas Street (NE, NW)
Stephen A. Douglas (1813–1861), Democratic congressman and senator, ran for reelection against Abraham Lincoln in 1858. The two locked political horns in the now-famous Lincoln-Douglas Debates. Nominated for president in 1860, Douglas ran second to Lincoln.

Formerly Detroit Street, the name was changed in 1905. The city's commissioners received many irate letters from residents concerning the name change. They mistakenly believed that the street would honor abolitionist Frederick Douglass. Many of the property owners stated that they would not continue to own or reside on a street bearing the name of a black man.

Downing Place (NE)
Andrew Jackson Downing was appointed by President Millard Fillmore to develop the Mall area. The Mall was an important part of Pierre L'Enfant's original plan, but in the 1850s it was an eyesore, highlighted by overgrown foliage and a stinking Washington Canal.

Duddington Place (SE)
Duddington Place is located near the site of Pierre L'Enfant's destructive attempt to see his plan for Washington become reality (see New Jersey Avenue). After L'Enfant tore down the walls of Daniel Carroll's house, Carroll decided to rebuild his house, Duddington. Duddington Place marks the spot today. His house was bounded by South Carolina Avenue, 2nd Street and E Street.

Dumbarton Rock Court, Street (NW)
Ninian Beall was captured while fighting for Charles II at the Battle of Dunbar. He was sent first to Barbados, then to Maryland, where he was freed. Upon his release, he received 50 acres of land.

Beall became known as a great political and military figure. Aside from his mill duties, he held many political offices as well as the title of Commander-in-Chief of the Provincial Forces of Maryland. For his public service and bravery, Beall was awarded approximately 25,000 acres of land, much of which has become national landmarks. The 1,503 acres called "Inclosure" became part of the National Arboretum, the White House now stands on the 255-acre "Beall's Levels" and "Rock of Dumbarton" is now preserved as the historic landmark "Dumbarton Oaks."

Dumbarton Oaks, named by Ninian Beall for the seaport and county town of Dumbartonshire, Scotland, has held many names over the years, depending upon the tastes of the owners. Almost as many historic events have taken place on the land. While owned by Charles Carroll of Bellevue, the land was appropriately called Bellevue. During the British attack of 1814, the White House was burned; Carroll rescued First Lady Dolly Madison. She took refuge on this property. In 1944, Dumbarton Oaks was the site of the Dumbarton Oaks Conference, which set the stage for the establishment of the United Nations.

Dunbar Road (SE)

Dunbar Road is named for poet Paul Lawrence Dunbar (1872–1906). Dunbar, the son of former slaves, earned his living operating an elevator. He wrote poetry in his spare time; his poems often appeared in newspapers. Dunbar wrote of Southern black life on plantations, using dialects from the area. After receiving a favorable review from William Dean Howells, Dunbar was sought for many events, including a reading engagement in England in 1897. Dunbar's most famous works include the school song for the Tuskegee Institute.

Dupont Circle (NW), Street (SE)

Admiral Samuel Francis duPont (1803–1865) was drawn to naval service by the achievements of the navy during the War of 1812. During the 1850s, duPont helped organize the Naval Academy and promoted military use of steam power. In 1860, duPont served as commandant of the Philadelphia Navy Yard. In 1861, he was appointed to the Commission of Conference, whose duties included planning naval strategy for the Civil War.

During the Civil War, duPont served as commander of the South Atlantic Blockading Squadron. In a substantial loss to the Confederates at Charleston, duPont publicly declared that the monitors (armored battleships) had been ineffective and were unfit for battle. He demanded that Lincoln release this information to the public; Lincoln remained silent and duPont asked to be relieved of his command, feeling that Lincoln's silence was an attempt to discredit him.

A statue of duPont was erected on Pacific Circle in 1884 and the circle's name was changed. The statue did not remain, though. The duPont family moved it to Wilmington, Delaware, in 1921 and commissioned Daniel Chester French to design a fountain to replace it (see French Drive). The fountain, often surrounded by men playing chess, remains the focal point of the circle today.

In 1949, two tunnels were constructed beneath the circle. The tunnels alleviated congestion on the trolley line. In 1962, when trolley service ended in the District of Columbia, the tunnels were sealed off.

Dupont Circle in the late 1800s was one of the seats of Washington high society. Residents included James G. Blaine (see Blaine Street) and Christian Heurich. Heurich's mansion was the second fireproof building built in the city. (The first was his brewery near the Potomac River on 25th Street, NW.) Joseph West Moore described it: "Fashion has firmly set its seal upon this district, and all those improvements which come with opulence have been lavished upon it."

During a massive urban renewal project in the 1950s, which some termed "urban removal," all but two of the lavish mansions were razed. Sleek office buildings were built around the circle.

Eads Street (NE)

Eads Street is named for inventor James Buchanan Eads (1820–1887). In 1842, using his invention of a diving bell for locating sunken steamers, Eads began making money as part of a steamship salvaging operation.

During the Civil War, Eads was hard at work. First, he submitted a plan to Abraham Lincoln for the defense of the western rivers. After creating a strategic plan, Eads suggested the construction of a fleet of steam-powered artillery boats.

After the war, Eads set about doing the impossible. A team of engineers had stated that it would be impossible to construct a bridge over the Mississippi River at St. Louis, but Eads proved them wrong. The bridge, which took seven years to build, was completed in 1874. That same year, Eads proposed the clearing of the mouth of the Mississippi. At the time, sediment from the river collected instead of flowing out into the Gulf of Mexico, making the river impassable to ships. Once again the engineers scoffed, but Eads successfully redirected the river's flow using a jetty system. His success made New Orleans a major seaport. Port Eads, at the mouth of the river, is named in his honor. (See Benton Street.)

Eagle Nest Road (NE)
Located on the grounds of the National Arboretum, Eagle Nest Road pays tribute to just one variety of bird that has found the area suitable for residence.

East Capitol Street (NE)
East Capitol Street extends from the Capitol, crossing the Anacostia River, to the Maryland border. This street creates the division between the Northeast and Southeast sections of the District.

Eastern Avenue (NE)
Eastern Avenue marks the eastern border of the District of Columbia, running along the Maryland line from Silver Spring to Capitol Heights. The street was named in 1906. The section between Chestnut and Cedar Streets in the Takoma Park neighborhood was formerly named Magnolia Street.

East Executive Avenue (NW)
This street used to provide easy access to the Treasury Department from the White House. Increased security, though, has made the street accessible only to pedestrian traffic. The street was closed to vehicular traffic in 1983.

Eaton Place, Road (SE)
Tennessee Senator John Eaton (1790–1856) caused controversy in Washington society and Andrew Jackson's cabinet by his marriage to Margaret "Peggy" O'Neal.

O'Neal was married to John B. Timberlake, a Navy officer. Timberlake was often out to sea; O'Neal became good friends with Eaton. Their friendship became an affair and when Timberlake died, the two decided to marry. Their marriage created a chasm in Washington's high society as people began arguing whether it was proper for the two to marry or not. O'Neal's party invitations were declined and she was quickly shut out of her former social circles.

Andrew Jackson, a friend to both Eaton and O'Neal, supported the couple. In 1829, Jackson appointed Eaton secretary of war. The cabinet became divided: Half supported the president, half had wives who refused to allow them to socialize with Peggy. It is said that Jackson had to force his cabinet members to attend state dinners. The chasm had become too wide for Jackson to bridge; he decided to dissolve his cabinet. In 1831, Eaton resigned his position.

Eaton subsequently served as governor of Florida. His wife did not like the state, so she stayed in Washington. When her husband was appointed minister to Spain, though, Peggy hurried to his side.

When Eaton died in 1856, O'Neal received large sums of money and property. Soon thereafter, O'Neal met an Italian dance instructor, Antonio Buchignani. She rented property to him in return for dance lessons for her grandchildren. Quite a charmer, Buchignani soon moved into O'Neal's house. The couple were married in 1859. In 1865, Buchignani took off with all of O'Neal's money and one of her granddaughters, with whom he produced a child. Buchignani returned to Washington in 1867 and O'Neal had him arrested. She refused to consent to his release until their divorce was final and he agreed to marry her granddaughter.

Eckington Place (NE)
The summer home of Joseph Gales was named Eckington; it lends its name to this street. The Eckington and Soldier's Home Railway, which ran between those two locations, was the first electric railway in the District of Columbia. (See Gales Street.)

Edgevale Terrace (NW)
Edgevale Terrace is located on the edge of Rock Creek valley.

Edmunds Street (NW)
George F. Edmunds (1828–1919) was a motivated politician. A successful lawyer in Vermont, he turned his sights to public office in 1854. He served one term in the state legislature, three years of the term as speaker. In 1861, he held the office of president pro tem of the state senate.

Edmunds went to Washington in 1866 to fill a vacant Senate seat. He was re-elected at the appropriate times by his constituents until he retired in 1891. Initially associated with the Radical Republicans, Edmunds was extremely powerful in Congress. He pushed the Tenure of Office Act through in 1867, oversaw the rules and procedures for the impeachment trial of Andrew Johnson (he voted for impeachment), drafted the Civil Rights Act of 1875 that was later overturned by the Supreme Court and drafted the act establishing the Electoral Commission in 1877, which was organized to decide the election of 1876. He sponsored the Edmunds Act of 1882, which outlawed polygamy in the territories, and he contributed to the Sherman Anti-Trust Act of 1890.

Edmunds, a sarcastic and aloof man, served as Chairman of the Judiciary Committee for 16 years. He was nominated for President in 1880 and 1884 by the Republican party. Despite much support, Edmunds was not successful. He moved a heartbeat away from the White House when, after the death of James Garfield, he was appointed president pro tem of the Senate.

Eglin Way (SW)
Eglin Way, located on the grounds of Bolling Air Force Base, is named for Eglin Air Force Base.

Elder Street (NW)
See Flora and Fauna.

Ellicott Street (NW), Terrace (NW)
Andrew Ellicott, a veteran of the Revolutionary War, was assigned by George Washington to survey the ten-mile

square that would become Washington. Prior to this undertaking, Ellicott surveyed the western portion of the Mason-Dixon Line and some boundaries of New York and Pennsylvania. Ellicott was assisted in his surveys by astronomer Benjamin Banneker (see Banneker Drive).

Ellicott was asked to have the city plan engraved. Pierre L'Enfant had failed to complete the task, which was a necessary element in the sale of lots in the new capital. L'Enfant was infuriated by Ellicott's unauthorized use of his work: ". . . to my great surprise it now is, unmercifully spoiled and altered from the original plan to a degree indeed evidently tending to disgrace me."

Ellipse Road (NE, NW)

There are two Ellipse Roads in the District of Columbia. The first, located in the Northwest section, outlines the famous Ellipse positioned between the White House and the Washington Monument.

The second road is located on the grounds of the National Arboretum. It is named for its geometrical shape.

Elmira Street (SE)

See City Streets.

Elm Street (NE, NW)

See Flora and Fauna.

ELLIPSE ROAD. Aptly named, Ellipse Road forms the perimeter of the Ellipse. Sixteenth Street extends behind the White House. Fourteenth Street and Federal Triangle are on the right. The National Archives is located on the upper right side of the Ellipse. [CREDIT: ALA.]

Embassy Park Drive (NW)
Located in the Embassy Park Townhouse complex off of Massachusetts Avenue, this is a continuation of Macomb Street.

Emerson Street (NE, NW)
Ralph Waldo Emerson (1803–82) was an American author and thinker. Emerson resigned a Unitarian pastorate and travelled Europe. Upon his return to the United States, Emerson set up a center of transcendentalism in Concord, Massachusetts, with the help of Margaret Fuller and Henry David Thoreau. He is also noted for a poem that appears on the Concord monument commemorating the start of the Revolutionary War, "the shot heard 'round the world."

Erie Street (SE)
Erie Street is named for Lake Erie, the fourth largest of the Great Lakes. The lake served as the setting for a naval battle during the War of 1812. Lake Erie separates Pennsylvania from Canada.

Euclid Street (NW)
Euclid Street is named for the Greek mathmetician Euclid (fl. 300 B.C.). The Euclidean theories of plane geometry have made schoolchildren's heads swim.

Evarts Street (NE)
Formerly Emporia Street, the name was changed in 1905 to honor William M. Evarts (1818–1901), an active lawyer and politician. Evarts attracted much attention with his stance on slavery: He opposed slavery, but believed it was constitutionally legal. Evarts, who started his political career as a Whig, led the New York delegation to the Republican national convention.

Evarts exercised his legal talents often. In 1867, he prosecuted Jefferson Davis for treason and in 1868, he served as the chief defense counsel at Andrew Johnson's impeachment proceedings. Grateful, Johnson appointed

Evarts Street (NE)

Evarts attorney general shortly after the proceedings failed.

In 1870, Evarts founded, and served for five years as president of, the New York City Bar Association. A leading candidate for chief justice of the Supreme Court, Evarts was denied by opposition organized by Senator Roscoe Conkling. In 1877, he served as Secretary of State under Rutherford B. Hayes. He was elected to the Senate in 1885, but served only one term due to failing health.

Everett Street (NW)

Everett Street is named for Edward H. Everett (1794–1865). Everett, a clergyman, was elected to the House of Representatives in 1821. He served for 10 years. He held positions as governor of Massachusetts (1836–39), minister to England (1841–45), president of Harvard University (1846–1849) and Secretary of State (1852–53) under Millard Fillmore. Everett concluded his political career by serving one year in the Senate. He was forced to resign due to differences of opinion between himself and his Abolitionist constituents in Massachusetts.

During the Civil War, Everett travelled the country vigorously supporting the Union cause. On November 19, 1863, Everett delivered the principal speech at the Civil War cemetery dedication that produced Abraham Lincoln's famous Gettysburg Address. Everett, whose address lasted two hours, wrote to Lincoln the next day: "I should be glad if I could flatter myself that I came as near the central idea of the occasion in two hours as you did in two minutes."

Everett's most widely used, if not best known, contribution to American life is his invention of the crimped metal Coca-Cola bottle top.

Fairmont Street (NW)

Fairmont Street is another street that was embroiled in a vicious debate during the renaming of 1905. It was originally Yale Street, but many residents wanted the name changed to Franklin Street. They felt that the name Fairmont gave undeserved publicity to the Fairmont School located at the western end of the street. The commissioners denied the request, stating that the name Franklin had already been used in another part of Washington. Also, a continuation of Yale Street was already named Fairmont Street, so changing the name to an existing street name simplified matters.

Faraday Place (NE, NW)
Faraday Place is named for Michael Farraday (1791-1867), an English scientist who developed the first dynamo. His invention led to the development of today's generators.

Farragut Place (NE), Street (NW), Square (NW)
David Glasgow Farragut (1801-1870) was living in Norfolk when Virginia seceded. As a Union supporter, he was forced to move north. Farragut is most famous for shouting "Damn the torpedoes! Full steam ahead!" while tied to the rigging of the *Hartford* during the Battle of Mobile Bay. Farragut was the first U.S. Naval officer to receive the ranks of vice admiral (1864) and admiral (1866).

The statue of Farragut that overlooks the square was sculpted by Vinnie Ream from photos supplied to her by Farragut's wife. In 1861, the square served as a Union artillery camp.

Fenton Court (NE), Place (NW)
Reuben E. Fenton (1819-1885) served in the New York State Assembly and the House of Representatives as a Democrat, but the slavery issue forced him to abandon the party. He helped found the Republican party in New York and returned to Congress five years later. His political hold over the state was so strong that when he ran for governor of New York in 1864, he won a larger proportion of the state's votes than Lincoln did in his presidential reelection bid. As his power within the state began to slip, Fenton abandoned politics and established himself as a financial expert.

Fenwick Street (NE)
Fenwick Street is named for John Fenwick, one of the proprietors of New Jersey.

Fern Place, Street (NW)
See Flora and Fauna.

Fessenden Street (NW)
Known as ill-tempered and insensitive, William P. Fessenden (1806–1869) served as chairman of the Senate Finance Committee (1857) and Secretary of the Treasury (1864). Fessenden resigned his federal appointment a year later and returned to Congress, beating out Hannibal Hamlin, Lincoln's first vice president, for the seat. Fessenden strongly opposed impeaching Andrew Johnson, stating that the proceedings were brought about by the "general cussedness" of the Radical Republicans in Congress. His stance caused many within the Republican party to question his loyalty, but his vote kept Johnson in the White House.

Fifteenth Street
Once the home of banks, brokerage houses, and, of course, the Treasury Department, 15th Street was nicknamed "Little Wall Street." When the financial businesses moved to K Street, the nickname became obsolete.

Fitch Place (NE), Street (SE)
John Fitch (1743–1798) experimented with steam engines. He is credited with the invention of the first functional steamboat, although at the time most of the glory went to Robert Fulton for his more commercially recognized steamboat.

Flint Place (NE)
See City Streets.

Flora and Fauna
In 1901, the Commissioners of the District of Columbia realized that at some point the east/west streets in the city would extend beyond the three syllable mark. They decided to name the subsequent streets after members

Flora and Fauna

of the plant kingdom. These streets are thought to be what remains of 140 acres of nurseries tended to and owned by John Saul, an Irish horticulturist. Although many streets are named for flowers and trees, the following are the ones that conform to the plan set forth by Congress.

    Aspen Street
    Butternut Street
    Cedar Street
    Dahlia Street
    Elder Street
    Fern Street
    Floral Street
    Geranium Street
    Hemlock Street
    Holly Street
    Iris Street
    Jonquil Street
    Juniper Street
    Kalmia Road
    Locust Road
    Myrtle Street
    Primrose Road

Other streets named for members of the plant kingdom are:

    Apple Road
    Beech Street
    Berry Road
    Birch Street
    Buckeye Drive
    Cherry Road
    Chestnut Street
    Dogwood Street
    Elder Street
    Elm Street
    Forest Lane
    Greenleaf Street/Greenvale Street
    Ivy Street
    Laurel Street
    Linden Place
    Magnolia Street
    Maple Street/Mapleview Place
    Oak Street/Oakdale Place/Oakview Terrace/
      Oakwood Street

Olive Street
Orange Street
Orchid Street
Pecan Street
Poplar Street
Redbud Lane
Redwood Drive
Spruce Street
Sycamore Street
Verbena Street
Vine Street
Walnut Street
Willow Street

The following streets can be found on the grounds of the National Arboretum:
Azalea Road
Beechwood Road
Conifer Road
Crabtree Road
Holly Spring Road
Rhododendron Valley Road

Floral Street (NW), Terrace (NW)
See Flora and Fauna.

Florida Avenue (NE, NW)
Named Boundary Street until 1890, this thoroughfare served as one of L'Enfant's original boundary lines. Later, it became the road to Bladensburg, Maryland. It was named Florida Avenue because it was a wide diagonal street, although it does not adhere to the original plan of geographical location (it is southeast of Connecticut and north of Rhode Island). (See State Streets.)

Foote Street (NE)
A former senator and Mississippi governor, Henry S. Foote (1804–1880) served two terms in the Confederate Congress, despite his support of Unionism. In 1865, Foote became frustrated with the Confederate Congress's failed attempts at peace. He decided to head north to conduct his own talks with President Abraham

Lincoln, but Union soldiers captured him before he could cross the Potomac. Attempts to have him expelled from the Confederate Congress failed and several days later he again attempted to gain access to Washington. This time he succeeded, only to have Lincoln refuse to see him.

Forest Lane (NW)
See Flora and Fauna.

Fort Drive (NW), Place (SE)
Fort Drive provides access to Fort Reno, located off of Fessenden Street between Wisconsin and Nebraska Avenues. Fort Place is located on the grounds of Fort Stanton Park. Unlike Fort Reno, Fort Stanton no longer stands. Fort Stanton was located between today's Suitland Parkway and Good Hope Road.

Fort Baker Drive (SE)
Fort Baker was located along Ridge Road (now Alabama Avenue) along the southeastern border of the District. Fort Baker Drive marks the location of the fort today.

The fort was named for Edward Dickinson Baker (1811–1861). Baker's political career carried him across the country. He led the Republican party in California and Oregon and served as a representative from Illinois. Abraham Lincoln was impressed by him (he named one of his sons after Baker) and it was not long before Baker gained national recognition. In 1861, Baker, with hopes of advancing his political career, put together a Pennsylvania regiment. He led his troops across the Potomac in an attempt to attack a Confederate camp outside of Leesburg, Virginia. Baker's expedition proved that he was an inept commander. With no set strategy, no means of retreat and little knowledge of the Confederate force he was facing, Baker entered a no-win situation and lost his life and the lives of many others. Due to his popularity as a politician, Baker was not blamed for the defeat. Instead, Brigadier General Charles P. Stone, who thought the advance was foolish, had his career ruined. Baker became a national hero.

Fort Davis Place (SE)
Fort Davis Place is located south of the exact location of the fort, which was situated along Ridge Road (now Alabama Avenue).

Fort Dupont Drive (SE), Terrace (SE)
Fort Dupont Drive is located to the southwest of Fort Dupont Park, which now surrounds the original location of the Civil War fort. Fort Dupont, built in line with Forts Baker and Davis, helped fortify the southeastern border of the District. (See Dupont Circle.)

Fort Lincoln Drive (NE)
Fort Lincoln Drive is located near the site protected by the fort. On the northeastern boundary of the District, Fort Lincoln, which was named for Abraham Lincoln, was located north of Bladensburg Turnpike (now Bladensburg Road). The fort's main purpose was to protect the route to Baltimore and the railroad line. Both routes were necessary in the transport of supplies.

Fort Stevens Drive (NW)
Fort Stevens, located at 13th and Quakenbos Streets, was the sight of the only serious Confederate attack on the capital city. In July 1864, Lt. Jubal A. Early's troops clashed with Union troops at the fort. President Abraham Lincoln and his wife were visiting the fort at the time of the attack. Lincoln was so intrigued by the fighting that he had to be ordered from his open position on the parapet. The next day, Lincoln again visited the fort and came under fire. A young Union officer noted that Lincoln, allowing curiosity to get the better of him, was again leaving himself open to fire. The officer, Oliver Wendell Holmes, shouted, "Get down, you fool!" After two days of skirmishing, Early turned his troops away.

Fort Stevens, originally Fort Massachusetts, was renamed to honor Brigadier General Isaac Ingalls Stevens (1818–1862). Stevens was a sickly child, but this did not keep him from an illustrious military career. He graduated first in his class from West Point. As a member of the Corps of Engineers, he assisted in the construction of fortifications along the Eastern coast. Stevens served in the Mexican War under Winfield Scott, then entered the political arena as an ardent supporter of Franklin Pierce. He resigned his military commission in 1853 when he was appointed governor of the Washington Territory. While travelling to his new job, Stevens directed the survey of the northern route of the transcontinental railroad. Stevens lost a bit of popularity in the Northwest when he displayed dictatorial tendencies during negotiations with the Indians. At the outbreak of the Civil War, Stevens returned to the military. He saw battle at Secessionville and the second battle of Bull Run. At Chantilly, Virginia, Stevens's troops were attacked by "Stonewall" Jackson. In the heat of the battle, Stevens was killed by a rifle ball.

Fort Totten Drive (NE)

Named for General Joseph Gilbert Totten (1788–1864), Fort Totten was located near Rock Creek Church Road along the northeastern boundary of the District. Still in existence, Fort Totten has become the focus of a park as well as a neighborhood.

Totten was a member of West Point's inaugural class; he later served as its first professor of mathematics. In 1806, Totten resigned from the military to help survey the Northwest Territory. Two years later, he rejoined the Army Corps of Engineers. Totten served as Winfield Scott's chief engineer during the Mexican War. He was considered a premier engineer. His work included constructing the fortifications in New York City, assisting with coastal fortifications and improving river and harbor conditions to ease civil transport. Totten, as a member of the Lighthouse Board, helped establish a lighting system along the Atlantic coast that did much to alleviate navigational hazards. Totten also used his engineering expertise as chief engineer of a commission appointed by Secretary of War Stanton to study Washington's defenses.

Foxhall Road (NW)

Convinced by Thomas Jefferson to move his iron foundry from Philadelphia in 1800, Henry Foxall became

Foxhall Road (NW)

America's first defense contractor. Foxall is credited with developing the cannon used by Commodore Oliver Hazard Perry at the Battle of Lake Erie. His foundry, Spring Hill, was located in Georgetown.

A former minister, Foxall had the Ten Commandments engraved on a marble slab and built into the foundry foundation. During the British attack on Washington in 1814, Foxall vowed to build a Methodist church in the city if the British spared his foundry. The British did, and the first church was built on the corner of 15th and G Streets. The church had to be relocated due to poor attendance and increased noise pollution. A new church was constructed in 1902 at Church and 16th Streets. Foxall also built churches at the Navy Yard and in Georgetown. The church in Georgetown was constructed for the black worshippers of the city. Foxall was also involved in Washington politics. He served as mayor of Georgetown from 1821–1823.

Foxall's name was misspelled on one of the first street signs. The spelling was never corrected.

Foxall's house was located at 34th Street on the C&O Canal, a few blocks from the road named in his honor.

Frankfort Street (SE)

See City Streets.

Franklin Street (NE, NW)

Formerly Frankfort Street, this street was renamed in 1905 to honor Benjamin Franklin (1706–1790).

Probably best known for his electricity experiment involving a kite, a key and bolt of lightning, Franklin is considered a great statesman. Franklin arrived in Philadelphia in 1723 and quickly became famous for his *Poor Richard's Almanack.* While still in Philadelphia, Franklin helped establish today's University of Pennsylvania. At the Albany Congress in 1754, Franklin proposed a union plan for the new colonies. He travelled to England, serving as an agent for several colonies, and returned to America in 1775. Shortly after his return he assisted in the drafting of the Declaration of Independence, which he signed. In 1781, Franklin was appointed to negotiate peace with Britain. In 1787, in one of his last acts as a public figure, he attended the Federal Constitutional Convention.

Frederick Douglass Court (NE)
Abolitionist Frederick Douglass (1817–1895) was known as "The Sage of Anacostia." English friends purchased his freedom in 1847 and he established the *North Star* in Rochester, NY to promote his stance on abolition. During the Civil War, he urged blacks to join the Union cause. In addition to his duties as newspaper editor and lecturer, Douglass served as Secretary of the Santo Domingo Commission (1871), marshall of the District of Columbia (1877–81), Recorder of Deeds for D.C. (1881–86) and minister to Haiti (1889–91).

French Drive (NW)
French Drive, which extends from the southeastern side of the Lincoln Memorial to Independence Avenue, is named for the sculptor of the statue inside. Daniel Chester French, using his knowledge of sign language, portrayed Lincoln's hands signing his initials (left hand "A," right hand "L").

Fulton Place (NE), Street (NW), Terrace (NW)
An artist, engineer and inventor, Robert Fulton (1765–1815) built and operated the *Clermont* on the Hudson River, the first successful industrial use of the steamboat. Fulton, guided by his American success, attempted to convince the French of the feasibility of steamboats along the Seine . . . they didn't buy it.

Gainesville Street (SE)
See City Streets.

Galen Street (SE)
Greek physician and writer Galen (c. 120–c. 200), through animal dissections, contributed greatly to med-

ical knowledge about the brain, nervous system, spinal cord and arteries. Unfortunately, his word was considered gospel; very few experiments by other people were undertaken until the 16th century, thus hampering medical progress.

Galena Place (NW)
Galena is a sulphide of lead, the principal source of the metal. This ore was first mined in Maryland in 1813.

Gales Place, Street (NE)
From 1807 to 1820 Joseph Gales, the only reporter covering Senate proceedings for the *National Intelligencer*, enjoyed the privilege of sitting next to the Vice President. In 1810, Gales purchased the newspaper and, two years later, shared the business with his brother-in-law William Seaton. As a public service, the two men published transcripts of Senate proceedings (a precursor to the *Congressional Record*).

Born in England, Gales was the object of Admiral George Cockburn's animosity when the British attacked Washington in 1814. Cockburn had Gales's papers burned and, legend says, ordered all c's at the newspaper destroyed to prevent his name from appearing in the publication. Gales went on to serve as mayor of Washington from 1827–1830. (See Seaton Place.)

Gallatin Place (NE), Street (NE, NW)
As Secretary of the Treasury (1801–14), Albert Gallatin (1761–1849) used Jeffersonian ideals to change U.S. financial policies. He advocated the use of federal money to stimulate the economy. Gallatin was a primary negotiator for the Treaty of Ghent, which ended the War of 1812. He also served as minister to France (1816–23) and Britain (1826–27).

Gallaudet Street (NE)
Gallaudet Street, named for Thomas H. Gallaudet, runs northeast of the university named in his honor.

A friend of the family witnessed Gallaudet attempting to teach words to the man's daughter, a deaf-mute. The father, with the assistance of other acquaintances, sent Gallaudet to Europe to become educated in formal methods of teaching deaf-mutes.

Upon his return to the United States in 1817, he opened a school for deaf-mutes in Hartford, Connecticut. The Connecticut Asylum, later renamed the American Asylum, was the first school of its kind in the country. It was also the first special school to receive state funds.

Gallaudet's son, Edward (1837-1917), joined Amos Kendall in the establishment of the Columbia Institution for the Deaf and Dumb, now Gallaudet University. (See Kendall Street.)

Prior to the recent appointment of I. King Jordan, the university had never been governed by a deaf president. His appointment was prompted by vigorous student demonstration that closed the university for several days.

Galloway Street (NE)

Joseph Galloway (c. 1731-1803) believed in changing the state government of Pennsylvania from a proprietary to a royal form. A prominent lawyer, Galloway served in the Pennsylvania assembly from 1757 to 1774. He was the speaker for the last eight years of his tenure. He was elected to the first Continental Congress in 1774. He proposed Galloway's Plan of Union, which was an attempt to solve the problem of home rule among the colonies. The plan was accepted early, but rejected in a later vote. All references to Galloway's plan were removed from the official records. Due to this defeat, Galloway refused to serve a term in the next Congress and spoke against independence.

Extremely unpopular, Galloway spent much time at his country home, where his friend Benjamin Franklin attempted to change his Loyalist views. Galloway advanced with British General William Howe through New Jersey in 1775; he commanded the British civil government during the British occupation (1777-1778). He returned to England with the British forces in 1776. Later he publicly criticized Howe and sought a reconciliation with the colonies. He firmly believed that America would profit from a tie to Britain.

In 1788, the Pennsylvania Assembly ordered the sale of Galloway's property after accusing him of high treason. In 1793, he placed a request to return to the state, but it was rejected. He spent the rest of his life in England, where he wrote several books about the conflicts between the Americans and British.

Galveston Place (SE), Street (SE)
See City Streets.

Garfield Street (NW), Terrace (NW)
James Abram Garfield (1831–1881), the 20th president of the United States, served from March to September 1881. He was shot on July 2 by a disgruntled office seeker, Charles J. Guiteau, and died on September 19. Garfield was succeeded by Chester A. Arthur (1830–1886), who ran his administration as the cog of a political machine.

Garrison Street (NW)
William Lloyd Garrison (1805–1879) campaigned for the abolition of slavery through the American Anti-Slavery Society (1833) and the founding of the *Liberator* (1831). While opposing the Civil War, Garrison urged the Northern states to secede from the Union also to protest the legality of slavery in the Constitution. He changed his stance when the Emancipation Proclamation was issued. During the nineteenth century, Garrison was considered one of the preeminent antislavery leaders, but now it is thought that other abolitionists were more effective.

Gates Road (NW)
Gates Road is named for Revolutionary War General Horatio Gates (1727–1806). Gates is best known as the hero of the Saratoga Campaign in 1777. His military career ended with a disgraceful defeat of the Patriot forces at Camden, South Carolina, in 1780.

Georgia Avenue (NW)
See State Streets.

Geranium Court (NW), Street (NW)
See Flora and Fauna.

Giesboro Road (SW)
Giesboro Road runs south for a half mile, terminating at the Maryland border. The street bears the name of a house built in that area during the seventeenth century. The patent for the land, the first one issued in present-day Washington, went to Thomas Dent on September 4, 1663. At that time, the property name was Gisbrough.

The Anacostia property on which Giesborough stood was owned by two families, the Addisons and the Youngs, from 1685 to 1921. Joseph Young described the estate in 1840 as "one of the finest around Washington and . . . a model farm in all respects. It was here that McCormick tested his first reaper." Giesboro, as it was also spelled, was one of several homes borrowed by the federal government for use as offices during the Civil War. This headquarters for Union officers had an estimated $37,000 damage at the war's end.

The house was burned in 1931 to clear land for the construction of Bolling Air Force Base.

Glover Drive, Road (NW)
Charles Carroll Glover (1780–1827) was intent upon improving the District of Columbia. A financier and civic leader, Glover gave land and money to create area parks. Glover donated 80 acres to the city for the creation of Glover-Archbold Park and two tracts of land to enlarge Fort Dupont Park in Anacostia. In the early 1920s, developers attempted to obtain land in Rock Creek Park. Glover gave his time and money to stop their efforts and in 1924, the park was saved by the creation of the National Capital Park Commission.

Glover also believed in art education and enjoyment. He built the Corcoran Gallery and served as its president from 1906 to 1933. (See Corcoran Street.)

Good Hope Road (SE)
This street's claim to fame: John Wilkes Booth used it as an escape route after assassinating President Abraham Lincoln.

Grant Circle, Road (NW)
The 18th president of the United States, Ulysses S. Grant (1822–1885), assisted in the naming of Washington streets. (See Tunlaw Road.) Grant, who served as commander of the Union Army during the Civil War, was the first U.S. citizen after George Washington to hold the rank of general.

Grant Street (NE)
In all the commotion of the renaming of streets in 1905, Grant Street was changed, without authority, to Pleasant Street. Residents became confused when street signs were posted bearing the "new" name, but the new maps still listed the original name. A surveyor eventually confirmed that Grant Street was the name recorded in the official records.

Greene Place (NW)
Greene Place is named for General Nathanael Greene (1742–1786), the Revolutionary War officer who ranked second to George Washington.

Greenleaf Street (SW)
See Flora and Fauna.

Greenvale Street (NW)
See Flora and Fauna.

Greenwich Parkway (NW)
See City Streets.

Gresham Place (NW)
Walter Q. Gresham held several cabinet positions. He served as Postmaster General (1883–84) and Secretary of the Treasury (1884) under Chester A. Arthur and as Secretary of State (1893–95) under Grover Cleveland.

Hadfield Lane (NW)
George Hadfield was hired by John Van Ness in 1832 to build a tomb for his wife, Marcia Burnes Van Ness. The tomb, located in Oak Hill Cemetery, imitated the Roman Temple of Vesta.

Half Street (SE, SW)
Located west of South Capitol Street, Half Street, SW, runs from Buzzard's Point on the Anacostia River to I Street. Half Street, SE, is located east of South Capitol Street, running between Potomac Avenue and I Street. In 1905, it was discovered that streets had been created running parallel to South Capitol Street and the two First Streets. Since the point of the 1905 renaming procedure was to make the nomenclature consistent with L'Enfant's plan and the guidelines of the commissioners, this development presented a problem. The commissioners had to either rename all the numbered streets east and west of South Capitol Street or give the streets names not in keeping with the plan. The commissioners opted for a compromise and Half Street, in keeping with the spirit of the numbering system, was created.

Hall Place (NW)
Astronomer Asaph Hall (1829–1907), after finding he could not make much money at "star-gazing," took a position as professor of mathematics at the Naval Observatory. Using the 26-inch telescope (which is still in use at the facility), he made many solar observations, including those relating to the motion of the planets. His greatest discovery came in 1877 when he discovered the two moons of Mars. Hall named the moons Phobos and Deimos (Fear and Anxiety) and carefully charted their movements to determine their orbits.

Hall Place is located to the west of the Naval Observatory.

Halley Street, Terrace (SE)
This street is named for astronomer Edmund Halley (1656–1742). Halley was the first to predict the orbit of a comet (the one that now bears his name). His study used Isaac Newton's theory of gravity.

Hamilton Street (NE, NW)
Alexander Hamilton (1755–1804) served as George Washington's secretary and aide-de-camp during the American Revolution. He served in the Continental Congress, giving his support to a capital city in the South, and as a delegate (New York) to the Federal Constitutional Convention. As Secretary of the Treasury under Washington (1789–95) he sponsored legislation to pay off the debts incurred by the Continental Congress (Assumption Act) and to charter the Bank of the United States. He was killed in 1804 by Aaron Burr in a duel; he had stymied Burr's bids for the presidency and the governorship of New York. Incidentally, a few years earlier, Burr's son was killed in a duel in the same place.

HAMILTON CIRCLE. The circle, located at the intersection of Massachusetts and Idaho Avenues, was eliminated by the construction of Christ Church. [CREDIT: ALA]

Hamilton Circle
Hamilton Circle, named in honor of Alexander Hamilton, once brought Massachusetts and Idaho Avenues together. The circle was removed in 1967 by the construction of Christ Church. City planners felt that there would not be enough room for both the church and the circle.

Hamlin Place (NE), Street (NE)
Hannibal Hamlin (1809–1891) served as Abraham Lincoln's first vice president, a position he did not want. During his vice presidency, Hamlin found himself too far from the combat action. To remedy this situation, Hamlin enlisted in the Maine Coast Guard and was criticized by many. By 1864, Hamlin had warmed to the second seat and expected to be renominated, but Lincoln chose Andrew Johnson. (See Fessenden Street.)

Hanna Place (SE)
Marcus Alonzo (Mark) Hanna (1837–1904) managed William McKinley's 1896 presidential campaign and served as a Republican senator from Ohio from 1897 until his death in 1904.

Harewood Road (NE, NW)
Harewood, William Wilson Corcoran's summer home, was seized by the government during the Civil War for office space. (He "lent" two buildings during the war; his newly constructed art gallery met the same fate.) Corcoran's home, though, was so badly damaged by the government employees occupying it that he refused to reclaim it. The government no longer had a use for the house, so it became the property of the Soldiers' Home. Harewood Road, which separates the U.S. Soldiers' Home from the campus of Catholic University, is all that remains of Corcoran's residence. (See Corcoran Street.)

Harlan Place (NW)
James Harlan (1820–1899) never considered a career in politics until he became involved in the free-soil movement. He served two terms in the Senate, once as a Whig and again as a Republican. Lincoln appointed him Secretary of the Interior (not surprising since his daughter married Lincoln's oldest son), a position he held into the Johnson Administration. Harlan and Johnson did not see eye to eye politically and their differences forced Harlan's resignation. After his third term in the Senate, Harlan's political rivals prevented him from holding any elected offices.

Harrison Street (NW)
Harrison Street was named to honor the ninth and 23rd presidents of the United States: William Henry Harrison (1773–1841) and Benjamin Harrison (1833–1901).

William Henry Harrison served as president for one month before dying of pneumonia. He used his ties to the Indians by running with John Tyler under the slogan "Tippecanoe and Tyler, too." After his death,

Harrison Street (NW)

Harrison was replaced by Tyler, who was considered an "acting" president.

The grandson of W. H. Harrison, Benjamin was elected president by the electoral college even though opponent Grover Cleveland won the popular vote. The first Pan-American conference was held during his term.

Hartford Street (SE)
See City Streets.

Harvard Street (NW), Court (NW)
Harvard Street is named for Harvard University, one of five Ivy League streets in the Columbia Heights neighborhood.

Hawaii Avenue (NE)
See State Streets.

Hawthorne Drive (NE), Lane (NW), Place (NW), Street (NW)
Judging from the number of streets bearing his name, Nathaniel Hawthorne (1804–1864) was extremely popular in Washington, D.C. Hawthorne created a new standard in fiction; his tales often involved complex moral and spiritual predicaments. Some of his famous novels include *The Scarlet Letter* and *The House of the Seven Gables*. His short story collections brought the genre to popularity.

Hayes Street (NE)
Hayes Street is named for the 19th president, Rutherford B. Hayes (1822–1893). Hayes, though, might not have been president without the creation of the electoral college, which decided the election of 1876. His Democratic opponent was Samuel J. Tilden (see Tilden Street). As the votes were being tabulated, the returns from South Carolina, Louisiana, Florida and Oregon were

disputed. Congress created the electoral college to decide the matter. The commission gave all the disputed votes to Hayes and he won the election by a one-vote margin in the electoral college. The decision affected the way the Hayes administration was viewed by the American public.

Hemlock Street (NW)
See Flora and Fauna.

Henry Bacon Drive
See Bacon Drive.

Hoban Road (NW)
After Pierre L'Enfant failed to complete the design for the President's House, the commissioners decided to hold a competition. Thomas Jefferson submitted drawings under an alias, but the winning design belonged to Irishman James Hoban (1762–1831). Hoban had also entered the design competition for the Capitol, but he did not win. Hoban, who had travelled to Washington after completing the Capitol in Columbia, South Carolina, decided to stay. He helped build the president's mansion (soon known as the White House due to the Virginia freestone used in its construction) from 1793 to 1800 and rebuilt it after the British burned it in 1814. He designed and built the offices of the state and war departments as well as Blodgett's Hotel. He also supervised the architectural design of the Capitol. He became active in Washington politics as a member of the City Council, where he served from 1820 until his death.

Hobart Place (NW), Street (NW)
Garret A. Hobart (1844–1899) served as vice-president of the United States under William McKinley. Hobart practiced law in New Jersey, but he soon developed an interest in politics. He was elected to the state legislature in 1873 and again in 1877, serving a total of seven years. He served 11 years as chairman of the New Jersey Republican committee; in 1884 he became a member of the

Hobart Place (NW), Street (NW)

Republican National Committee. Hobart never served in Congress, but his popularity and stance on pertinent election issues earned him the vice-presidential nomination. He died after two years in office.

Holly Street (NW, SE)
See Flora and Fauna.

Holly Spring Road (NE)
See Flora and Fauna.

Holmead Place (NW)
James Holmead, Jr., owned a large plot of land that would become upper Northwest Washington. In 1740, Anthony Holmead built a log cabin around the 3500 block of 13th Street (parallel to Holmead Place), which expanded to Holmead Manor. Ten years later, James Holmead built his home, Rock Hill, near S and 24th Streets today. Rock Hill was later renamed Kalorama and served as a place for the Holmeads to supervise work at their mill on Rock Creek. (See Kalorama Road.)

Hopkins Street (NW)
Hopkins Street is named for philanthropist Johns Hopkins (1795–1873). A branch of Johns Hopkins University is located one block away.

Hospital Road (NW)
Hospital Road is located on the grounds of the U.S. Soldier's Home.

Howard Place (NW), Road (SE), Street (NW)
Howard University was founded to provide education to emancipated slaves by an act of Congress in 1867. Funds for the creation of the university were provided by the Freedmen's Bureau. General Oliver O. Howard, a white

Union officer, managed the Bureau after the Civil War. He faced a court of inquiry to answer charges of mismanagement. Despite these charges, Howard served as the university's president from 1869 to 1874.

Hurst Terrace (NW)
John Fletcher Hurst was a Methodist bishop. In 1890, Hurst chose the site for the American University at the intersection of Massachusetts Avenue and Loughboro Road (now Nebraska Avenue). Hurst Hall, the university's first building, was built in 1896.

Idaho Avenue (NW)
See State Streets.

Illinois Avenue (NW)
See State Streets.

Independence Avenue (SE, SW)
Independence Avenue runs parallel to Constitution Avenue. The two streets form the border for the Mall. In 1931, the street was named. Since the Constitution ensured independence, the two streets were considered appropriately placed running past the monuments and buildings of the federal government of the United States.

INDEPENDENCE AVENUE. Independence Avenue as it travels behind the Mall (left) and beside the Department of Agriculture. [CREDIT: ALA.]

Indiana Avenue (NW)
The original Indiana Avenue ceased to exist with the construction of government buildings around 1907. Louisiana Avenue, which now faced the buildings, was renamed Indiana Avenue. Louisiana moved east to Union Station. (See State Streets.)

Ingleside Terrace (NW)
Located in Mount Pleasant, Ingleside was the first farm in one of the oldest Washington neighborhoods. The house on the property was built for a close friend of George Washington, Harry Ingle.

Ingraham Street (NE, NW)
Joseph Holt Ingraham (1809–1860) began his career writing sensational, violent novels. After reaching a self-proclaimed total of 80 novels, Ingraham opened a school for girls in Nashville, Tennessee. By 1851, something overcame Ingraham and he underwent a religious rebirth. In 1852 he was ordained an Episcopal priest.

Ingraham's religious transformation did not halt his writing. He produced three biblical novels, all of which were extremely popular. Ingraham used a share of the royalties from these novels to buy up copies of his earlier works. He considered his previous writing an embarrassment to his new life. Ingraham spent the rest of his life in parishes in Mississippi, Alabama and Tennessee.

Iowa Avenue (NW)
See State Streets.

Iris Street (NW)
See Flora and Fauna.

Irving Street (NE, NW, SE)
Formerly Concord Street, the street was renamed in 1904 to honor Washington Irving (1783–1859). Perhaps best known as the author of "Rip Van Winkle" and "The

Legend of Sleepy Hollow" and George Washington's biography, Irving also served as minister to Spain (1826–29).

Ivy Street (SE)
See Flora and Fauna.

Jackson Street (NE), Place (NW)
Formerly Dover Street, this street was renamed in 1904 to honor Old Hickory, Andrew Jackson (1767–1845), the seventh president of the United States. Spurred by his popularity in the West, Jackson almost beat John Quincy Adams for the highest office in 1824. The decision was made in the House of Representatives and Adams became president. When he took office in 1828, Jackson became known for his political stances and his powerful Kitchen Cabinet. The spoils system, which rewarded those loyal to the current party in power with political appointments, developed during Jackson's administration.

Jackson fought to abolish the Bank of the United States and the issue became the focal point of the 1832 election. The election pitted Jackson against Henry Clay. After his reelection, Jackson removed all federal money from the bank and placed it in state banks.

Jamaica Street (NE)
Jamaica Street is named for the West Indies republic. Jamaica became independent in 1962.

Jay Street (NE)
Jay Street is named for the first chief justice of the Supreme Court, John Jay (1745–1829). In 1773, he served as secretary of the New York–New Jersey boundary commission. The following year, he served in the Continental Congress. Although he did not support independence at first, Jay threw his support into the cause as it gained speed. He worked feverishly for the ratification of the Declaration of Independence. In 1776, he helped create the New York state constitution and served as that state's Chief Justice for three years. Jay was reelected to

Congress in 1778 and served as president for nine months. He relinquished his office when he was named minister to Spain in 1779.

Jay, a supporter of a strong Federal government, wrote five of the Federalist Papers. He became Chief Justice in 1789, while serving as Secretary of State until Thomas Jefferson could be sworn in. During his five year occupation of the top seat the procedures for the Supreme Court were outlined. In 1794, Jay negotiated a treaty, subsequently named for him, to handle the payment of pre-Revolutionary war debts to British merchants and reimbursement for confiscated property.

After being defeated by George Clinton in 1792, Jay was elected governor of New York in 1795. Jay, who was in England when elected, served two terms.

Jefferson Drive (SW), Place (NW), Street (NE, NW)

Thomas Jefferson (1743–1826), who helped plan the city of Washington, served as the third president of the United States. He tied Aaron Burr in votes, but he was elected to office by the House of Representatives. Author of the Declaration of Independence, Jefferson was the first president inaugurated in Washington. Jefferson's administration concentrated mainly on foreign affairs. He oversaw the Louisiana Purchase (1803) and planned the Lewis and Clark expedition (1803–1806), which explored the land of the Louisiana Purchase west

JEFFERSON DRIVE. Thomas Jefferson, one of the city planners, is immortalized in the Jefferson Memorial. The memorial, located along the Tidal Basin, is on land that was resurrected from the Potomac River in 1882. [CREDIT: Washington Convention and Visitors Bureau.]

to the Pacific Ocean. After retiring in Virginia, Jefferson helped found the University of Virginia around 1809.

Jenifer Street (NW)
Jenifer Street is named for Daniel Jenifer of St. Thomas, a signer of the Constitution from Maryland.

John Marr Circle (SE)
John Q. Marr (1825-1861) was the first Confederate soldier to die in combat. He was killed by a random bullet fired by the 2nd U.S. Cavalry as they rode through Fairfax Court House, Virginia.

John Marshall Place
This street, named for the fourth chief justice of the Supreme Court, used to run between Pennsylvania Avenue and D Street, NW. The street was destroyed by the construction of Judiciary Square. John Marshall Place, formerly 4½ Street, was removed in the late 1970s. Between the two World Wars, 4½ was diminished to 4th Street. The street was the dividing line between black and white neighborhoods. In this segregated time, it came as quite a shock when these neighbors joined forces to successfully lobby the city for street improvements.

John Marshall (1755-1835), a strong supporter of the Constitution, served as John Adams' secretary of state. Once appointed chief justice, Marshall began setting constitutional precedents. The case of *Marbury v. Madison* brought the Supreme Court the power to determine the legality of legislation. He determined the Constitutional strength of the federal government in *McCulloch v. Maryland* and *Gibbons v. Ogden*. Marshall's legal strides while on the high court brought him the distinction of the "Great Chief Justice."

Johnson Avenue (NW)
Andrew Johnson (1808-1875) served as the 17th president of the U.S. As a senator representing Tennessee (1857-1862), Johnson championed Southern causes un-

til his state seceded; he then became a staunch Lincoln ally. He was appointed military governor of Tennessee in 1862. As a Southerner and War Democrat, he provided considerable balance to Lincoln's bid for reelection. He gained the presidency after Lincoln's assassination.

Johnson gained no support from Radical Republicans and faced much opposition to his Reconstruction plans. He became the only president against whom Congress has held impeachment proceedings. The impeachment, brought about by his attempts to remove Edwin Stanton from his position as Secretary of War, failed to pass by one vote.

Joliet Street (SE)
One of several Joliet Streets in existence in the District of Columbia at the turn of this century. (See City Streets.)

Jonquil Street (NW)
See Flora and Fauna.

Juniper Street (NW)
See Flora and Fauna.

Justice Court (NE)
Justice Court is named for its location: one block north of the U.S. Supreme Court.

Kalmia Road (NW)
The kalmia is an evergreen shrub, developed by Swedish botanist Peter Kalm. (See Flora and Fauna.)

Kalorama Road (NW)
Calorama is Greek for beautiful or fine view, which is exactly what Joel Barlow (1754–1812) had from his suburban estate. The land, purchased in 1805, sat on Rock Creek approximately one mile from the District boundary at that time. Barlow died in Poland in 1812, but the area he developed became Washington's first suburb.

In 1887, Joel Barlow published a popular 12-volume epic poem "The Columbiad" (originally titled "The Vision of Columbus"), which exalted the area. His best-known poetical work, though, is "The Hasty-Pudding."

Barlow served as consul in Algiers, negotiating treaties and the release of American prisoners. Named Citizen of France for his work supporting the French Revolution, he joined the ranks of George Washington, Alexander Hamilton, James Madison and Thomas Paine as the only Americans to receive the honor. Barlow was also a financial and physical backer of Robert Fulton's steamship experiments. Legend states that Barlow allowed Fulton to test his ship in the section of Rock Creek near his house. Barlow died of exposure in Zarnowiec, Poland, caught up in Napoleon's retreat from Moscow.

Kalorama, which was used as a smallpox hospital during the Civil War, burned during a farewell ball for the troops. The building was removed in 1889.

Kalorama Road, named for the Barlow estate, does not fit into the alphabetical scheme of Washington street names because the commissioners felt it was "in a section of the District where streets are at such angles it is impossible to make them conform to the general system."

Kanawha Street (NW)

As well as being a city in West Virginia, Kanawha is a county and river in that state. (See City Streets.)

Kansas Avenue (NE, NW)

See State Streets.

Kendall Street (NE)

As a journalist and editor of the *Argus of Western America*, Amos Kendall (1789–1869) was instrumental in the development of Andrew Jackson's political power. He served as a member of Jackson's Kitchen Cabinet and as postmaster general under Jackson and Martin Van Buren (1835–40). During his postal tenure, Kendall, in support of Jackson, confiscated abolitionist William Lloyd Garrison's mail from railroad trains. Kendall acquired a fortune by acting as Samuel F. B. Morse's agent for distribution and use of the telegraph patents. His greatest accomplishment, though, was the creation of Gallaudet University.

Kendall founded a school for the deaf, which he named after himself, on his estate at 7th Street and Florida Avenue, NE (where Gallaudet University now stands). The school's name was changed to the Columbia Institution for the Deaf, Dumb and Blind and Kendall lobbied Congress to incorporate the school. They complied in 1857 and Kendall asked Edward Gallaudet, who had experience working with the deaf, to run the school. Kendall put $20,000 toward the school's operation. The school, which became the National Deaf Mute College, was authorized by Congress to grant degrees in 1864. Gallaudet ran the college for 53 years, and after his death, the university was named for his father, a pioneer in the field. (See Gallaudet Street.)

Kendall also financed the Calvary Baptist Church, 8th and H Streets, NW.

Kensington Place (NE)
See City Streets.

Kentucky Avenue (SE)
See State Streets.

Kenyon Street (NW)
Kenyon Street is one of four remaining streets named for American universities. It is named for Kenyon College in Gambier, Ohio.

King Place (NW)
Rufus King (1755–1827) served as a delegate to the Continental Congress (1784–1787) where he helped draft the Ordinance of 1787, which provided for the creation and government of the Northwest Territory. King served as U.S. senator from New York (1789–1796, 1813–1825) and, after each stint in the Senate, he served as minister to Great Britain (1796–1803, 1825–1826).

Kirby Street (NW)
Edmund Kirby's (1840–1863) family was split by the Civil War; he chose to fight for the Union. Kirby saw action in most of the major eastern campaigns until he was wounded at Chancellorsville. He was sent to a hospital in Washington, but did not survive. He was promoted to brigadier general on the day of his death.

Klingle Road (NW)
Klingle Road travels through the Cleveland Park neighborhood from the National Cathedral to the eastern border of Rock Creek Park. In 1961, a representative from Tennessee attempted to have the street renamed Tennessee Avenue. The reason: The existing four-block street did not do the state justice.

Knox Circle (SE), Place (SE), Street (SE), Terrace (SE)
Henry Knox (1750–1806), a hero of the Revolutionary War, had an early interest in the military. At 18, he joined the militia. He served under General Artemas Ward (see Ward Circle) as a volunteer during the Battle of Bunker Hill and at the commencement of the Boston Seige. Knox's "Noble Train of Artillery," which moved heavy artillery over 300 miles of rough roads to Boston, was directly responsible for the British retreat from that city. Knox set up defenses at vulnerable points in Connecticut and Rhode Island. Knox and his artillery forces were powerful during the Battle of Long Island, as well as during battles in New York and New Jersey. He established the Springfield arsenal in Massachusetts in 1777, while serving as chief of the artillery.

Described as "forceful, often profane," with a "pompous, self-complacent walk," the 300-pound Knox commanded the fort at West Point from 1782–1784 and succeeded George Washington as commander-in-chief from 1783–1784. Knox was elected Secretary of War in 1785 and held the office for nine years, although he did not officially begin his term until 1789. Knox died as a result of consuming a chicken bone. Fort Knox is named in his honor.

Laboratory Road (SW)

Laboratory Road is located on the grounds of Bolling Air Force Base. It provides access to the Naval Research Laboratory.

Lackland Way (SW)

Lackland Way, located on Bolling Air Force Base, is named for Lackland Air Force Base.

Lafayette Avenue (NE)

Marie Joseph Paul Yves Roch Gilbert du Motier, Marquis de Lafayette (1757–1834), better known as just Marquis de Lafayette, joined the American cause despite the disapproval of his family and the King of France. Lafayette carried papers from Silas Deane (see Deane Avenue), commissioning him as a major general in the Continental Army. In 1777, when Lafayette reached Philadelphia, he agreed to pay his own expenses and start as a volunteer. At 19, Lafayette was eager to serve, although he spoke only a few words of English and had never seen or fought in battle before. At Brandywine (1777), Lafayette helped halt the British advance. This show of bravery earned him command of the Virginia light troops in December 1777. A staunch supporter of George Washington, Lafayette served at Valley Forge and commanded a light division in the last battle against Cornwallis.

Lafayette spent approximately $200,000 supporting the American Revolution. In 1794, despite Lafayette's lack of interest in restitution, Congress paid him approximately $24,500. In 1803, Congress gave him 11,500 acres in Louisiana.

In 1784, Washington invited Lafayette to return to America. He provided assistance concerning political and economic matters to Thomas Jefferson during his tenure as U.S. minister to France.

In 1789, Lafayette took command of the new French National Guard. In 1792, he led the 52,000-man Army of the Center into battle when war against Austria was declared. His command was removed due to a shift in political power and, after fleeing to Belgium, he was captured and imprisoned for five years. Freed by Napoleon, Lafayette refused offers of political positions. He also refused Thomas Jefferson's offer of governorship of Louisiana in 1805. He removed himself from political life.

In 1824, Lafayette returned to the United States at the invitation of President James Monroe. His arrival at Staten Island was greeted by "demonstrations of frenzied enthusiasm without precedent or parallel in American history." Lafayette's influence on American history is felt not only in Washington, but throughout the cities of the United States, i.e. 13 named Fayette, six named Fayetteville and 14 named Lafayette.

Lafayette Square (NW)

This square, located across the street from the White House, was fittingly named President's Square. During a year-long "triumphal return" to the United States by the Marquis de Lafayette (1757–1834) in 1824, the square was renamed in his honor as citizens of Washington thronged to the square to see him. The focal point of the square, though, is a statue of Andrew Jackson. The statue, dedicated in 1853, was the first equestrian statue erected in the United States. The statue is constructed of brass from guns captured by Jackson during the War of 1812 at the Battle of Pensacola. The four guns located at the base of the monument were captured by Jackson at the Battle of New Orleans.

This area became a real estate development for Washington's elite. Commodore Stephen Decatur was the first to build a private house on the land. Benjamin Ogle Tayloe lived in the Octagon and Dolly Madison resided in the Cutts-Madison House (given to her to repay a debt incurred by her brother-in-law, Congressman Richard Cutts). Today, Lafayette Square teems with protesters attempting to catch the president's attention.

Lamont Street (NW)

Formerly named Dartmouth Street, in honor of Dartmouth College in Hanover, New Hampshire, the street name was changed during the street reorganization of 1905. The present name honors Daniel S. Lamont, Secretary of War (1893–97) under Grover Cleveland.

Langley Court (NE)

Dr. Samuel Langley served as the secretary of the Smithsonian. In the 1880s, after returning from the West with bison, he established the Department of Living Animals, which was part of the National Museum. The collection, initially prompted by his interest in taxidermy, was located on the mall. The collection was later moved to its present location and renamed the National Zoological Park.

Langley also conducted other scientific experiments. He was convinced that heavier-than-air machines could fly and he tested his theories many times. In 1896, he launched a steam-powered aircraft from a catapult on the Potomac River. The craft flew several thousand feet. He tested a gas powered vessel in 1903; it flew a little further than the 1896 experiment. Langley, though, was unable to decisively prove his aircraft theory; he gave up due to lack of funds and the ridicule of his colleagues.

Langley Way (SW)

Langley Way, located on Bolling Air Force Base, is named for Langley Air Force Base in southeastern Virginia.

Lanier Place (NW)

Sidney Lanier (1842–1881) was fascinated by the union of poetry and music. His study of the subject was *The Science of English Verse*. His poetic works include "Corn" and "Song of the Chattahoochee."

Laurel Street (NW)

See Flora and Fauna.

Lawrence Avenue (NE), Lawrence Street (NE)

These two streets, approximately eight blocks apart, are named in honor of Commodore James Lawrence (1781–1813). Commander of the *Chesapeake* during the War of 1812, Lawrence was mortally wounded when defeated by the British frigate *Shannon*. His dying words, "Don't give up the ship," have become well-known.

Lawrence Street was named Hartford Street until 1905.

Lee Street (NE)

Lee Street is named for Confederate General Robert E. Lee (1807–1870). Lee, a veteran of the Mexican War, served three years as superintendent of West Point (1852–1855). Lee declined an appointment in the U.S. army when the South seceded, but accepted command of the Army of Northern Virginia (1862) when that state left the Union. Lee guided his troops into battle with an uncanny leadership ability. He fought in the Seven Days Battles, the second Battle of Bull Run, the Antietam Campaign, the Battle of Fredericksburg, the Battle of Chancellorsville, the Gettysburg Campaign and the Wilderness Campaign. Lee was appointed general-in-chief of all Confederate army forces in February 1865. He could not catapult the south to victory; he surrendered to U. S. Grant at Appomattox Courthouse two months later. After the war, Lee served as president of Washington College (presently Washington and Lee University).

L'Enfant Promenade, Plaza (SW)

Pierre Charles L'Enfant (1754–1825) came from France in 1777 and joined the Continental Army. He had engineering experience and was quickly commissioned as a major in the Corps of Engineers. In 1781, he was taken prisoner at the Siege of Charleston. His release was negotiated by General Rochambeau. His military service caught George Washington's attention and L'Enfant was promoted to major in 1782.

During his military service, L'Enfant gained a reputation as an artist. He spent much time sketching his fellow officers and Washington.

After the Revolutionary War, L'Enfant designed the insignia for the Society of the Cincinnati, which was founded by commissioned officers who served in the American Army and Navy during the war. When Congress moved from Philadelphia to New York, L'Enfant was asked to remodel New York City Hall to accommodate Congress. His design utilized symbols that have become uniquely American. It included medallions embossed with "U.S.," stars and stripes and an eagle with 13 arrows (signifying the number of states in the Union) emerging from its talons.

In 1789, "Monsieur Lanfang," as Washington called him, learned of Congress' intention to build a capital. He offered his services:

> . . . Your Excellency will not be surprised that my ambition and the desire I have of becoming a useful citizen should lead me to wish to share in the undertaking . . . No nation had ever before them the opportunity offered them of deliberately deciding on the spot where their Capital City should be fixed . . . And although the means now within the power of the country are not such as to pursue the design to any great extent, it will be obvious that the plan should be drawn on such a scale as to leave room for that aggrandizement and embellishment which the increase of the wealth of the nation will permit it to pursue at any period however remote.

Washington accepted L'Enfant's offer and asked him to design the new federal city. L'Enfant's design, which he completed in less than a year, included buildings and monuments that he thought reflected the grandeur of the new capital. He also planned parks and fountains along the thoroughfares of the city. Washington loved his ideas and accepted the plan.

Construction of the federal city began. Although L'Enfant has been described as "a man of many accomplishments, with an overflow of ideas and few competitors . . . the factotum of the new nation," he had an incredible temper and did not like to concede any detail to those he felt possessed less knowledge than himself. Therefore, he was in constant conflict with the commissioners. He felt that his plan should be implemented exactly as it appeared on paper. He flew into a rage when he discovered that Daniel Carroll, nephew of the commissioner of the same name, deliberately built his house in the middle of an avenue (New Jersey Avenue). L'Enfant, despite orders from Washington to cease, tore the walls of the house down. This act, combined with his desire to take part in the financial aspects of planning while neglecting his assigned duties, forced his dismissal from the commission.

L'Enfant's name was omitted from the original map. He floated from one job to another, dying virtually unnoticed in Prince George's County, Maryland. His grave was marked by a solitary tree. His accomplishments were not recognized until 1889. On April 28, 1908 his body was reinterred at Arlington National Cemetery, with a stunning view of the city he designed.

Lexington Place (NE)
See City Streets.

Library Court (SE)
Library Court is located on the fringes of the triangle formed by the three buildings of the Library of Congress: Jefferson, John Adams and James Madison.

Lincoln Drive (NW), Road (NE)
These two streets are named for Abraham Lincoln (1809–1865), the 16th president of the United States. Lincoln's political career began in 1834 when he was elected to the Illinois legislature. He served one term (1847–1849) as a Whig. Six years later his bid for a Senate seat was quashed. He joined the newly formed Republican Party and ran against Stephen A. Douglas in 1858 for a Senate seat (see Douglas Street). The campaign

LINCOLN DRIVE. Abraham Lincoln is honored with the Lincoln Memorial. The memorial is located at the west end of the Reflecting Pool. Constitution Avenue, formerly the Washington Canal, can be seen on the left. The Arlington Memorial Bridge is also visible on the left. (The draw is visible in the bridge's center span.) The Theodore Roosevelt Bridge crosses the Potomac River and Roosevelt Island on the right. The far shore is Arlington, Virginia. [CREDIT: ALA.]

produced seven famous debates. He lost the election, but gained political recognition. In 1860, he was nominated for president. Running against a divided Democratic party, he won. Southerners opposed his election and states began seceding from the Union in anticipation of his inauguration.

In 1863, Lincoln issued the Emancipation Proclamation, a move to free the slaves. That same year, he gave the Gettysburg Address. (See Everett Street.) In 1864, Lincoln won re-election. His victory is credited mainly to increased Union successes.

On Good Friday, April 14, 1865, Abraham Lincoln was assassinated by John Wilkes Booth while attending a performance at Ford's Theater.

Linnean Avenue (NW)

Linnean Avenue is named for Joshua Peirce's house, Linnean Hill. Peirce, a horticulturist, named the house overlooking Porter Street for Swedish botanist Carolus Linnaeus (1707–1778). Linnaeus is considered the creator of the modern system for classification of plants and animals. The house, built in 1823, became part of Rock Creek Park in 1890.

Peirce, who was also a nursery owner, provided trees for many of Washington's streets and Lafayette Park.

Little Falls Street (NW)

Little Falls Street, located on the perimeter of the Dalecarlia Reservoir, is named for the slightly rough section of the Potomac River adjacent to it.

Livingston Road (SE), Street (NW), Terrace (SE)

The Livingston family produced a number of American political figures:

Robert R. Livingston (1718–1775) was a judge on the New York Supreme Court and a delegate to the Stamp Act Congress.

Robert R. Livingston (1746–1813), son of the above, conducted the negotiations for the Louisiana Purchase. He also financed Robert Fulton's steamboat experiments.

Logan Circle (NW)

Edward Livingston (1764–1836) served in Congress (1795–1801) and as mayor of New York City. He relocated to Louisiana and served as a congressman (1823–1829) and senator (1829–1831). He also served as secretary of state under Andrew Jackson.

Philip Livingston (1716–1778) signed the Declaration of Independence and was an active participant in the Continental Congress. He was one of the founders of King's College (now Columbia University).

William Livingston (1723–1790) was the first governor of New Jersey (1776–1790). He also attended the first two Continental Congresses.

Locust Road (NW)

See Flora and Fauna.

Logan Circle (NW)

Logan Circle, originally named Iowa Circle, draws Rhode Island and Vermont Avenues together with 13th Street. This circle, named for General John Logan, was considered special by Pierre L'Enfant.

John A. Logan (1826–1886) served several terms in the Kentucky legislature and two terms in the House of Representatives. At the commencement of the Civil War, Logan's Union loyalties were questioned due to his family's relocation to Illinois. He supported laws barring free blacks from the state, thus his name was often used to bolster Confederate enlistment. (His brother-in-law joined the Confederate army.) In 1861, Logan gave a rousing speech to Union recruits and removed all doubt of his loyalty. His distinguished service with the Army of the Tennessee won him the support of U. S. Grant and Major General William T. Sherman. Logan commanded the Army of the Tennessee after Major General James B. McPherson was killed. He developed a resentment toward career soldiers, though, when Sherman replaced him with a West Point graduate.

After the war, Logan served terms in the House (1867–1871) and the Senate (1871–1877, 1879–1886). In 1884, he ran an unsuccessful campaign for vice president. Logan was a staunch ally to U. S. Grant, but he became known for advancing his career through corrup-

Logan Circle (NW)

tion. Logan, who suggested the establishment of Memorial Day, often used the holiday as a political stepping-stone.

L'Enfant's special circle was a residential area between 1872 and 1880, but as the areas further west became more sought after by home owners, Logan Circle became neglected. Many of the old Victorian homes were destroyed in the name of commercial progress. In the late 1960s the area was redeveloped and the circle now sports what has been described as "the most complete Victorian facade in the city."

Longfellow Street (NW)

Longfellow Street is named for poet William Wadsworth Longfellow (1807–1882). Longfellow is known for building American legends in narrative poems. His most famous works include *The Song of Hiawatha, Paul Revere's Ride* and *Evangeline.*

The commissioners renamed the street, formerly named Flint Street, in 1905 to make it conform to the nomenclature guidelines for the city. The Brightwood Park Citizens' Association requested that the name of one section of the street be changed to Livingstone. The commissioners denied the request for two reasons. First, they stated that a Livingstone Street already existed in Anacostia. Two streets with the same name in different parts of the city would cause confusion (and defeat the purpose of what they were trying to accomplish). The commissioners also claimed that the name had been recorded in the city records. Since only one section of the street was to be changed, with the other blocks remaining Longfellow Street, an act of Congress would be required to change it.

Loring Way (SW)

Nicknamed "Old Blizzards" while fending off Grant's attack from atop cotton bales during the siege of Vicksburg, William W. Loring (1818–1886) fought gallantly for the Confederacy. Loring feuded with "Stonewall" Jackson over winter quarters for his troops, feeling that Jackson was leaving his troops exposed to Union attack. After ending his military service early in 1865, Loring travelled to Egypt. He spent 10 years there and returned to write about his experiences in *A Confederate Soldier in Egypt*.

Loughboro Road (NW)

Nathan Loughborough served as Comptroller of the Treasury under President John Adams. He firmly believed the slogan "No taxation without representation" and refused to pay his taxes. The courts eventually forced him to make payment.

Loughborough owned 250 acres in present-day upper northwest Washington (much of the land is now The American University). His country home, located on this property, was named Grasslands, but he also owned several houses on M Street in Georgetown. Milton, a house he built on property adjacent to Grasslands, still stands at the intersection of River and Little Falls Roads. The Loughborough Mill was located on this property. The mill was destroyed during the Civil War by a group of drunken sailors. A relative, Hamilton Loughborough, gave much of the land to emancipated slaves at the end of the Civil War.

Loughboro Road, a continuation of Nebraska Avenue, runs from Chain Bridge Road to Macarthur Boulevard. There is no documented reason for the spelling difference, but, apparently, spelling errors were not foreign to Loughborough. During his tenure as Comptroller, he temporarily changed the spelling of his name to "Luffborough" to alleviate confusion in mail delivery.

Louisiana Avenue (NW)

Originally, Louisiana Avenue was situated near Judiciary Square. With the construction of government buildings around 1907, Indiana Avenue was buried. Louisiana Avenue was renamed Indiana; Louisiana soon reappeared a few blocks away at Union Station. (See State Streets.)

Lover's Lane (walkway) (NW)

Lover's Lane runs through a section of Rock Creek Park from R Street to Massachusetts Avenue. The road, which has long been closed to vehicular traffic, earned a reputation for secret rendezvous in the nineteenth century. In 1900, the name became official.

Lowell Lane (NW), Street (NW)
Charles R. Lowell (1835–1864) was considered a superb combat officer during his Civil War service. While serving under Major General George McClellan, he was appointed to carry the captured Confederate battle flags back to Washington as a reward for his bravery at Antietam. He helped defend Fort Stevens against Confederate attack and was mortally wounded three months later by the same troops at Cedar Creek.

Luray Place (NW)
See City Streets.

Luzon Avenue (NW)
Travelling between Walter Reed Army Medical Center and Rock Creek Park, Luzon Avenue commemorates the World War II battles waged on the largest of the Philippine islands. Luzon was invaded by the Japanese in 1941. The Allied forces faced the Japanese on the island at Bataan and the offshore island of Corregidor in 1942.

MacArthur Boulevard (NW)
Water from the Georgetown Reservoir is still conducted to the residents of Washington through the original pipes under Conduit Road. The road was built in 1863. The street name was changed in 1942 to honor General Douglas MacArthur (1880–1964), who was serving in the Pacific. This street is the last street in the District of Columbia to be named for a living person. (A law now states that streets may be named only for individuals who have been deceased a minimum of two years.)

MacArthur's promise "I shall return" echoed throughout the country during World War II, and his military career was marked by such heroics. During World War I, MacArthur organized and commanded the

42nd "Rainbow" Division. At the termination of the war, he was given the command of the U.S. occupation zone.

MacArthur proved that he was just as good in the field as he was in the office. He served in many administrative positions: superintendent of West Point (1919-1922), commander of the Philippines Department (1928-1930) and army chief of staff (1930-1935).

In 1935, MacArthur was assigned to the soon-to-be-independent Philippines to organize its defense forces. The army decided to transfer him to a new assignment two years later, but MacArthur did not feel his first task was complete. He resigned from the army and stayed in the Philippines. In 1941, with the threat of war in the Pacific becoming more serious, the Philippine and U.S. armies were merged. MacArthur was recalled to active duty and given command of this united force. On December 7, 1941, the Japanese attacked the Philippines as well as Pearl Harbor. MacArthur's troops were overwhelmed and pushed back to the Bataan Peninsula and, eventually, the island of Corregidor. MacArthur was ordered to depart the Philippines in 1942; at this time he uttered his famous promise. He was promoted and placed in command of the Allied forces in the Southwest Pacific arena. MacArthur began an island-hopping counteroffensive that helped to secure the Philippines in 1945. In October 1944, he fulfilled his promise by returning to the country. On September 2, 1945, MacArthur accepted the Japanese surrender.

Following the war, MacArthur was placed in command of the Allied occupation forces in Japan, where he assisted that country with economic and political organization.

In 1950, MacArthur was appointed supreme commander of U.N. forces in Korea. With the South Korean army and U.N. forces virtually defeated, MacArthur launched an amphibious attack against the North Koreans. MacArthur was relieved of his command due to publicly disagreeing with President Harry S. Truman's war policies. MacArthur's return to the United States, sparked by his heroic service in three wars, was greeted by an adoring public.

On a scenic note: The hills that run along this street were formed by the receding river waters.

MacArthur Terrace (NW)
See MacArthur Boulevard.

Macomb Street (NW)
Macomb Street is named for Alexander Macomb, a member of the Washington Monument Commission. The commission was charged with raising funds for the completion of the monument.

Madison Court (NW), Drive (SW), Place (NW), Street (NE, NW)
Madison Street is named for the fourth president of the United States, James Madison (1751–1836). Madison helped draft the Virginia Constitution. He served in the Continental Congress (1780–83, 1787) and in the Virginia legislature (1784–86). He is considered one of the creators of the Constitution and his contributions to the *Federalist Papers* facilitated the ratification of the Constitution in Virginia. While in Congress (1789–1797), Madison strongly supported the Bill of Rights. He served as Thomas Jefferson's secretary of state and followed him to the presidency in 1809. During his administration, he fought "Mr. Madison's War," the War of 1812.

MACOMB STREET. Glover-Archbold Park and the National Hebrew Congregation are located on this street at the corner of Massachusetts Avenue. [CREDIT: ALA.]

Magazine Road (SW)
Magazine Road is located on Bolling Air Force Base and provides access to the Naval Research Laboratory.

Magnolia Street (SE)
See Flora and Fauna.

Maine Avenue (SW)
See State Streets.

Manning Place (NW)
Daniel Manning served as Grover Cleveland's Secretary of the Treasury (1885–87).

Maple Street (NW)
See Flora and Fauna.

Mapleview Place (SE)
See Flora and Fauna.

Marine Place (NE)
No, Marine Place is not named for a branch of the U.S. military. It is so named for its location just east of the Kenilworth Aquatic Gardens.

Marion Street (NW)
Francis Marion (c. 1732–1795) earned his nickname, the "Swamp Fox," for his use of guerilla warfare during Revolutionary War battles with the British in South Carolina.

Marne Place (NE)
Two World War I battles were fought at the Marne River in France. The first, in 1914, aborted the German advance toward Paris. The second, in 1918, halted the

last significant German advance. This street, located near the Civil War's Fort Mahan, is named in honor of those battles.

Martin Luther King Jr. Avenue (SE)

This street runs across Anacostia from South Capitol Street to the 11th Street Bridge. The street, which bisects St. Elizabeth's Hospital, was originally named Asylum Avenue. The street was later renamed Nichols Avenue to honor St. E's superintendent, Charles Henry Nichols. The name was changed once again shortly after the civil rights leader's assassination.

Martin Luther King, Jr. (1929–1968), was a Baptist minister who gained national recognition for his support of passive resistance to segregation and his boycott of the segregated bus lines of Montgomery, Alabama. King's Southern Christian Leadership Conference provided a base for the 1963 March on Washington and the 1965 Selma, Alabama, voter registration drive. King was killed in Memphis by James Earl Ray.

Maryland Avenue (NE, SW)

See State Streets.

Massachusetts Avenue (NW, SE)

Massachusetts Avenue, home of Embassy Row and the U.S. Naval Observatory, is the longest avenue in the District of Columbia. (See State Streets.)

Embassy Row was, for the most part, created in the mid-1930s. The Great Depression forced many property owners to sell their mansions. The only ones with the money to buy these houses were foreign governments. Representatives for many governments purchased mansions, which they promptly converted into residences and offices for their ambassadors assigned to Washington.

In the early 1900s, the corner of Massachusetts and Wisconsin Avenues was thought to offer one of the best views of the city. James Bryce, a British ambassador

(1906–1913), described the view in a 1913 *National Geographic* article: "You all know the spot . . . just opposite where the Episcopal Cathedral is to stand. At that point, you look down upon the city, you see its most striking buildings . . . and beyond them the great silvery flood of the Potomac." The cathedral and apartment buildings now stand on the corner. Bryce's view can still be seen, but only from the roof of one of those buildings. Bryce's favorite scenic spot is marked, though. A small, triangular park across from the cathedral is named in his honor.

McGuire Avenue (SW)

Hunter H. McGuire (1835–1900) served as the chief medical officer of the Army of the Shenandoah and "Stonewall" Jackson's friend and physician. He stated that his main function was to handle all of Jackson's "exotic physical complaints and theories." Although greatly affected by Jackson's death due to complications from the arm amputation he had performed, McGuire went on to organize the Richmond University College of Medicine.

McKinley Place (NW), Street (NW)

McKinley Street is named for the 25th president of the United States, William McKinley (1843–1901). McKinley served in Congress (1877–91), but his strong support of protective tariffs cost him his congressional seat. He was elected governor of Ohio (1891, 1893) and elected president in 1896. Both political prizes were the result of support given by Marcus A. Hanna, an Ohio political boss. (See Hanna Place.) During McKinley's administration, the Spanish-American War was fought, which allowed the United States to emerge as a world power. McKinley set in place the highest tariff rate in American history, annexed Hawaii, established the Open Door policy in China and established the Currency Act of 1900, which consolidated the gold standard. McKinley was re-elected in 1900, but he was assassinated by Leon Czolgosz in Buffalo, New York, nine months into his second term.

Meade Street (NE)

George Gordon Meade (1815–1872) was appointed commander of the Army of the Potomac in 1863. A brave commander, Meade advanced at Antietam until his troops ran out of ammunition. After the Civil War, he served as the top general under Ulysses S. Grant. Meade's training, though, was as an engineer. He helped build many coastal lighthouses and Fort Pennsylvania in the District of Columbia.

Meadow Road (NE)

Meadow Road is located on the grounds of the National Arboretum.

Meigs Place (NE)

Engineer Montgomery C. Meigs was asked to devise a system to provide water to Washington. In 1852, the West Point graduate designed an aqueduct system that carried water from Great Falls. The project, which took over 10 years to complete, included the construction of the Union Arch Bridge. (See American Legion Bridge.)

Georgia-born, Meigs loathed Southerners who left the Union, including his own brother. Appointed by his close friend and associate Jefferson Davis to construct the aqueduct system, Meigs honored him by placing his name on the bridge. At the start of the Civil War, though, Meigs had Davis's name removed.

Meigs was considered strong and, sometimes, pushy. In the late 1850s, he was appointed supervising engineer for the Capitol. He ignored architect Thomas U. Walter and attempted to make the building project his own. He hired European craftsmen to add glitzy ornamentation, which was considered tasteless and un-American by many. He was removed from the project in 1861.

Meigs left his mark all over Washington, both literally and figuratively. (Meigs had his name chiselled onto the 10 miles of pipes comprising the aqueduct system and imprinted on the iron trusses of the Capitol.) He designed most of the 68 forts constructed during the Civil War and established Arlington National Cemetery. Many, though, consider the Pension Building, located at

4th and F Streets, his greatest, or at least most unique, architectural achievement.

The building, which now houses the National Building Museum, was constructed to honor the veterans of the Civil War. A frieze, depicting all aspects of the Union Army at war, takes a quarter mile lap around the building's exterior. Today it is considered a unique use of space, but the Pension Building was ridiculed by many architects when it was erected in 1883. The red brick and orange terra cotta are the first suggestion that the building breaks the stuffy mold of federal office buildings. The offices inside are entered from balconies overlooking a tremendous courtyard. Eight huge Corinthian columns, the largest ever constructed in the Roman style, grace the expanse, of which one architect said, "Nothing short of an inaugural ball or a thunderstorm could possibly fill that immense void." General William Tecumseh Sherman's criticism illustrated his concern: "The worst of it is, it is fireproof."

Mellon Street (SE)

Andrew W. Mellon (1855–1937) found his niche as Secretary of the Treasury. In 1886, Mellon, with his brother Richard, took over his father's banking business and expanded it. He invested in oil, coal, steel, public utilities and insurance. In 1921, he resigned as president of Mellon National Bank to become Secretary of the

MELLON STREET. Banking magnate Andrew V. Mellon served as secretary of the treasury under three presidents. Mellon provided the funds for the construction of the National Gallery of Art; he donated his art collection to the public for display at the gallery. [CREDIT: Washington Convention and Visitors Bureau.]

Mellon Street (SE)

Treasury, a position he held under three presidents: Warren Harding (1921-24), Calvin Coolidge (1924-29) and Herbert Hoover (1929-32). Mellon also served as ambassador to Great Britain (1931-32). In 1937, Mellon gave his art collection to the public and set forth funds for the construction of the National Gallery of Art.

Michigan Avenue (NE, NW)
See State Streets.

Military Road (NW)
Military Road today stretches from the Maryland border through Rock Creek Park to Missouri Avenue. The road was constructed during the Civil War to aid in the transport of food, grain and ammunition to and from Washington's forts. The road originally provided a route from Fort Sumner (near Chain Bridge) to Fort Stevens (on Georgia Avenue). During the war, wagon trains often stretched for miles along the road.

Mill Road (NW)
Mill Road, due to its close proximity to Rock Creek, is probably named for a mill that stood nearby.

Mills Avenue (NE)
Andrew Jackson appointed Robert Mills (1781-1855) "Federal Architect of Public Buildings," a position he held for 15 years. Mills left his mark on Washington, with his designs for the Patent Office, the Old Post Office, the Treasury Building and the Washington Monument. The majority of Mills's buildings met a fiery fate. The Patent Office and the Old Post Office burned in 1835, the Treasury Building burned in 1833. The Washington Monument seen on the Mall today is not the same one seen in Mills's plans. Mills envisioned the obelisk seated atop a columned base.

Minnesota Avenue (NE, SE)
See State Streets.

Mississippi Avenue (SE)
See State Streets.

Missouri Avenue (NW)
See State Streets.

Monroe Street (NE, NW)
James Monroe (1758-1831), the fifth president of the United States, was a veteran of the American Revolution. He served in the Virginia legislature and the Continental Congress, where he opposed the Constitution because he felt it would create a highly centralized government. He later served as governor of Virginia (1799-1802). In 1811, travelling to England, Spain and France, Monroe helped negotiate the Louisiana Purchase. He served as fellow Virginian James Madison's secretary of state (1811-17), and, for a brief period, as his secretary of war (1814-15). Monroe was elected president in 1816 and re-elected in 1820. His administration is noted for the Missouri Compromise, boundary settlements with Canada and the acquisition of Florida. The Monroe Doctrine, issued in 1923, set a standard for U.S. foreign policy. The doctrine, which was never formally recognized in international law, declared that the United States was no longer available for European colonization and any attempts to colonize would be seen as unfavorable.

Montana Avenue (NE)
See State Streets.

Morgan Street (NW)
This street is named for General Daniel Morgan. Morgan was considered a hero of the Revolutionary War. One story states that he won a battle because his hemorrhoids forced him to stand down from his horse. Morgan also served in Congress.

Morningside Drive (NW)
Morningside Drive, a curving street between Alaska Avenue and Kalmia Road, is named for a suburb of

Edinburgh, Scotland. The area is known for the well-refined accents of its inhabitants.

Morris Place (NE), Road (SE)
Robert Morris (1734–1806), a signer of the Declaration of Independence, earned a reputation as the "financier of the Revolution" due to his success in collecting money for George Washington's army.

Morse Street (NE)
Morse Street is named for the inventor of the telegraph and the code used with it, Samuel F. B. Morse (1791–1872). Morse spent 12 years perfecting the electric telegraph, a task he completed in a cottage on Amos Kendall's estate (see Kendall Street). Morse demonstrated his invention to Congress in 1844, outlining its practical uses, and gained world recognition. In 1845, Morse opened the first telegraph office (7th Street between E and F Streets). The telegraph was linked to Baltimore. Morse also experimented with submarine cable telegraphy.

Mount Olivet Road (NE)
Mount Olivet Road leads to the cemetery of the same name.

Mount Pleasant Street (NW)
Mount Pleasant Street, located in the neighborhood of the same name, sits atop a hill overlooking Rock Creek Park. When the area was farmland, the view from the hill was pleasant, indeed.

Mozart Place (NW)
Mozart Place is named for Austrian composer Wolfgang Amadeus Mozart (1756–1791). Mozart began composing at age five and was performing concerts throughout Europe by his early teens. Mozart is known for operas such as *The Marriage of Figaro* and *Don Giovanni* as well as numerous symphonies.

Murdock Mill Road (NW)
Murdock Mill was located slightly southwest of Massachusetts Avenue and the road that bears its name. The property was owned by William D. C. Murdock (1806–1886).

Murdock's family was active in the Washington area, before and after the creation of the capital city. His relative William Murdock was the Maryland representative to the Stamp Act Congress of 1765. His father, George Murdock, was the first rector of the Rock Creek Parish.

Nebraska Avenue (NW)
See State Streets.

Nelson Place (SE)
Considered a hero of the American Revolution, Thomas Nelson signed the Declaration of Independence and served as governor of Virginia. During the Battle of Yorktown, Nelson pointed to his own house, considered the nicest in town, and directed the artillery to destroy it. He wished to prevent British General Charles Cornwallis from using it as a residence.

Nevada Avenue (NW)
See State Streets.

Newark Street (NW)
In 1881, President Grover Cleveland anonymously (he thought) purchased a summer house on Newark Street in an attempt to beat the heat of the White House. Cleveland's house was known by several names: Forest Hill, Oak View and Red Top (for the red roof). The residents banded together and named the neighborhood Cleveland Park to honor the president. (See City Streets.)

Newcomb Street (SE)
Newcomb Street is named for astronomer Simon Newcomb (1835–1909). Newcomb observed the orbits of six

planets and created tables predicting their motions. His study of the moon helped establish accurate lunar tables. His findings were adopted by observatories around the world.

New Hampshire Avenue (NW)
See State Streets.

New Jersey Avenue (NW)
New Jersey Avenue would lead to Pierre L'Enfant's dismissal during the building of Washington. Daniel Carroll and Notley Young, two extremely influential landowners, began building their houses along one of L'Enfant's wide avenues. Unfortunately, the sides of the houses encroached upon the street as L'Enfant had drawn it. Not known for his even temperament and tact, L'Enfant began tearing down the walls of Carroll's house.

New Jersey Avenue had not developed much by the time Congress moved into the city. A Connecticut congressman commented on the state of the avenues: ". . . not one was visible, unless we except a road with two buildings on each side of it, called the New Jersey Avenue." (See State Streets.)

New Mexico Avenue (NW)
See State Streets.

Newport Place (NW)
See City Streets.

Newton Place (NW), Street (NE, NW)
John Newton (1822–1895) was known as a superior commander during the Civil War. In 1862, as a member of the Corps of Engineers, he helped build many of the forts around Washington. Newton led troops into battle at South Mountain, Antietam and Fredricksburg. New-

ton was one of the officers who requested the dismissal of General Ambrose Burnside after the defeat at Fredericksburg. They did not get any satisfaction as Lincoln sided with Burnside. Newton continued his Union service at Chancellorsville and Gettysburg. After the war, Newton again joined the Corps of Engineers, becoming its commander in 1884.

In 1905, while the commissioners were adjusting the street names, the name was changed from Howard Avenue. A resident, W. N. Campbell, suggested changing the name of this street to Nugent Street. He stated that 75 streets already existed in U.S. cities bearing the name Newton; Nugent was present in only five cities. He claimed the change would prevent misdirected mail. Besides, he added, his middle name was Nugent.

New York Avenue (NE, NW)
See State Streets.

Nicholson Avenue (NW, SE), Street (NE, NW)
These two streets are named for Governor Sir Francis Nicholson of Maryland.

North Capitol Street (NE, NW)
This street serves as the dividing line between the northwest and northeast quadrants of the city.

In 1906, a bill was presented in the Senate requesting that North Capitol Street be renamed Georgia Avenue. The senators sought the opinion of the commissioners; the response was unfavorable. The commissioners gave the following reasons for their dissent: 1) The name Georgia Avenue already existed in the southeast quadrant; 2) the name North Capitol Street was part of the original street plan for the city; 3) the street designates where numbers change from east/west to north/south and 4) since the street had been built up commercially and residentially, the expense required to change addresses would be immense. The commissioners recommended changing the name of Brightwood Avenue instead.

North Carolina Avenue (SE)
See State Streets.

North Dakota Avenue (NW)
See State Streets.

North Portal Drive (NW)
North Portal Drive is the northern boundary of the Montgomery Blair Portal, leading into Rock Creek Park.

Oak Drive (SE), Street (NW)
See Flora and Fauna.

Oakdale Place (NW)
Originally named Oak Street, a duplicate in the Washington street system, the commissioners decided to rename the street Oak Court in 1905. The residents contested that Oak Court was a "stupid name" and requested that the name be changed to Oak Place. At the time, a street with the same name existed and the commissioners denied their request. The residents persisted and submitted the name Oakdale Place. The commissioners, in January 1906, approved the request because the difference in names was not considerable.

Oklahoma Avenue (NE)
See State Streets.

Oakwood Street (SE), Terrace (NW)
See Flora and Fauna.

Observatory Circle (NW), Lane (NW), Place (NW)
The first Naval observatory was built near 23rd and D Streets, near where the Lincoln Memorial stands today.

Unfortunately, astronomical observation at the site was nearly impossible due to the swampy conditions and increasing traffic vibrations. Rutherford B. Hayes appointed a commission to choose an appropriate site "upon an even degree of longitude west of Greenwich, and to possess . . . clearness of atmosphere, freedom from obstruction from the horizon, and freedom from objectionable vibrations from traffic."

In 1844, the commission discovered a site in Georgetown Heights that fit the specifications perfectly and purchased 72 acres of Mrs. M. C. Barber's property, Pretty Prospect (see Prospect Street). Ten years passed, though, before the government appropriated funds to construct the observatory. At the same time, the District Department of Highways was planning to extend Massachusetts Avenue through the future observatory site.

The superintendent of the observatory quickly took action on his own: First, he began digging the basement of his house. Then he surveyed, from the top of the hill, a perimeter with a 1,000-foot radius. He considered this measurement a suitable distance to protect the sensitive instruments from vibrations. This area could not be encroached upon by the any other project, thus forming the arc of Observatory Circle as Massachusetts Avenue travels between Whitehaven and Calvert Streets.

The superintendent's house was completed in 1893, but the Navy commandeered it in 1928 for the official home of the Chief of Naval Operations. The official name became Admiral's House. In 1974, Congress, looking for a way to reduce the logistical pressure and expense of protecting the Vice President at his private home, designated the house the official residence of the Vice President. (Besides, Admiral Zumwalt complained that the roof leaked and the dining room was too small for parties of any social magnitude.) The house was refurbished and made available to Nelson Rockefeller, who chose to remain in his house on Foxhall Road. Walter Mondale became the first Vice Presidential resident.

As for the Observatory, it maintains data on three kinds of time: Universal Time (Greenwich Mean Time), Ephemeris Time and Atomic Time. The 100-year-old telescope used by Dr. Asaph Hall (see Hall Place) to discover the two moons of Mars is still in operation.

Oglethorpe Street (NE, NW)

Oglethorpe Street is named for James Oglethorpe (1696–1785). In 1728, Oglethorpe was a member of the House of Commons, sitting on a committee investigating penal conditions. Oglethorpe suggested creating a penal colony in America; he secured a charter in 1732. One year later, Oglethorpe accompanied 120 settlers to the colony of Georgia.

In 1736, Oglethorpe built Fort Frederica on St. Simons Island. The fort's main purpose was to repel possible Spanish attacks from the south. The fort served its purpose in 1742, three years after war had broken out between England and Spain. In 1752, after Oglethorpe surrendered the charter, Georgia became a royal province.

Ohio Drive (SW)

One of two state streets not designated an avenue. (See State Streets.)

Olive Street (NE, NW)

See Flora and Fauna.

Oneida Place (NW), Street (NE)

Oneida Street is named for a nation of the Iroquois Confederacy. The Oneida were the only members of the league who did not side with the British during the American Revolution. The Iroquois inhabited, as they do today, parts of New York and Ontario. A lake in north central New York is named for the Oneida.

Ontario Road (NW)

Ontario Road, located parallel to Champlain Street, is named for the smallest of the Great Lakes, Ontario. Lake Ontario separates New York from Canada.

Orange Street (SE)

See Flora and Fauna.

Orchid Street (NW)
See Flora and Fauna.

Ord Street (NE)
See Ordway Street.

Ordway Street (NW)
A favorite of Ulysses S. Grant, Edward Otho Cresap Ord (1818–1883) was considered an aggressive commander. Ord participated in John Brown's capture following his raid on Harpers Ferry. He fought in many battles during the Civil War, but in his last conflict, he helped destroy Robert E. Lee's Army of Northern Virginia.

Oregon Avenue (NW)
See State Streets.

Original Street Names
The following street names were changed or reassigned during the 1905 renaming. The renaming was designed to bring continuity to District street names and to make the street conform to Pierre L'Enfant's original plan.

Original Name	Became
Morgan Avenue	10th
Wallace	10th
Eslin Avenue	11th
Duncan	12th
Burns	13th
Argyle	14th
Columbia Avenue	15th
Trinidad	16th
Central Avenue	17th
Piney Branch Road	17th
Clagett	18th
Poplar	18th
19½ Place	20th
Columbus	20th

Original Street Names

Original Name	Became
Woodley Place	23rd
Lover's Lane	31st
Observatory Place	32nd
33rd	33rd Place
Hubbard Place	33rd Place
Folsom	34th Place
Wisconsin Avenue	37th
Richmond	38th
LeDroit Avenue	2nd
4th	3rd
Harewood Avenue	3rd
5th	4th
Linden	4th
6th	5th
Larch	5th
7th	6th
Juniper	6th
8th	7th
Wright's Road	8th
8th	9th
Queen	9th
Albany	Adams
Lanier Terrace	Adams Mill Road
Park Road	Adams Mill Road
Utica	Allison
Lyles Place	Alton Place
Albany	Ashby
Tahoe	Aspen
Grant	Barry Place
Grant Avenue	Barry Place
Prospect	Belmont
Staughton	Belmont
Belmont	Belmont Road
Baltimore	Biltmore
Allison	Brandywine
Baltimore	Bryant
Trumbull (east of 4th)	Bryant
Vallejo	Buchanan
Umatilla	Butternut
Armes Place	Butterworth Place
Cincinnati	Calvert
S	Cambridge Place

Original Street Names

Original Name	Became
Carroll Avenue	Carroll
Woodley Road	Cathedral Avenue
Oak Avenue	Cedar
Vermillion	Cedar
Champlain Avenue	Champlain
Cincinnati	Channing
College	Chapin
Morris	Chapin
Buchanan	Chesapeake
South	Chesapeake
Chestnut Avenue	Chestnut
Hurst Place	Clark Place
Cincinnati	Cleveland
19th	Cliffbourne Place
Clinton Avenue	Clinton
Q	College Place
Columbia	Columbia Road
Steuben	Columbia Road
Columbia Road (from Florida to California)	Connecticut Avenue
Lawrence Place	Cottage Place
Crescent	Crescent Place
Wilmington	Crittenden
Detroit	Cushing Place
Wabash	Dahlia
Emporia	Dana Place
Des Moines	Davenport
Prospect	Davenport
Xenia	Decatur
Kearney	Dennison Place
Cincinnati	Douglas
Detroit	Douglas
Welling Place	Douglas
Montgomery	East Avenue
Frankfort	Edmunds Place
Erie	Ellicott
Hancock	Ellicott
Yuma	Emerson
Erie	Euclid
Euclid Place	Euclid
Irving	Euclid
Roanoke	Euclid

Original Street Names

Original Name	Became
Detroit	Evarts
Emporia	Evarts
Bismark	Fairmont
Huntington Place	Fairmont
Lincoln	Fairmont
Yale	Fairmont
Zanesville	Farragut
Flint	Fessenden
Grant	Fessenden
Emory	Fessenden Place
Sherman	Fessenden Place
Savannah	Fort Drive
Ridge	Foxhall Road
Emporia	Franklin
Frankfort	Franklin
Huron	Fuller
Frankfort	Fulton
Franklin	Fulton
Albemarle	Gallatin
Evarts Place	Garfield
Galveston	Garfield
Binney	Girard
Frankfort	Girard
Galena Place	Girard
Galveston	Girard
Princeton	Girard
Sumner	Girard
Colfax	Gresham Place
Dearborn Place	Gresham Place
Brandywine	Hamilton
Austin	Hamlin
Fulton	Hamlin
Galveston	Hamlin
Hartford	Hamlin
Bacon	Harvard
Hartford	Hawthorne
Highland Avenue	Highland Road
Morris	Hobart Place
Holmead Avenue	Holmead Place
Howard Avenue	Howard Place
Chesapeake	Ingraham
Flint	Ingraham

Original Street Names

Original Name	Became
Concord	Irving
Gerry	Irving
Girard	Irving
Hartford	Irving
Indianapolis	Irving
Kenesaw	Irving
Kenesaw Avenue	Irving
McClellan	Irving
McClellan Avenue	Irving
Wallach	Irving
Dover	Jackson
Hamlin	Jackson
Indianapolis	Jackson
Joliet	Jackson
Des Moines	Jefferson
Joliet	Jewett
Kalorama Avenue	Kalorama Road
Prescott	Kalorama Road
Prescott Place	Kalorama Road
Superior	Kalorama Road
Frankfort	Kearny
Irving	Kearny
Joliet	Kearny
Keokuk	Kearny
Erie	Kennedy
Huron	Kennedy
Hancock	Kenyon
Hancock Avenue	Kenyon
Marshall	Kenyon
Joliet	Kilbourne Place
Clark Place	King Place
Keokuk	Klingle
Dartmouth	Lamont
Farragut	Lamont
Grant	Lamont
Ludlow Avenue	Lamont
Lanier Avenue	Lanier Place
Laurel Avenue	Laurel
Hartford	Lawrence
Jackson	Lawrence
Keokuk	Lawrence
Lowell	Lawrence

105

Original Street Names

Original Name	Became
Flint	Longfellow
Hamilton	Longfellow
Itaska	Longfellow
Pierrepont Place	Lowell
Gennessee	Madison
Ingraham	Madison
Juniata	Madison
Maple Avenue	Maple
King	Meigs Place
Meridian Avenue	Meridian Place
Messmore	Messmore Place
Milwaukee	Michler
Lansing	Monroe
Lowell	Monroe
Lydecker Avenue	Monroe
Milwaukee	Monroe
Sheridan	Monroe
12th	Montello Avenue
M	Morse
Sheridan	Morton
Warren Place	Murdock Place
N	Neal
Loughboro Road	Nebraska Avenue
Rusk	New Cut Road
T	New Cut Road
Riggs Road	New Hampshire Avenue
Tunlaw Road	New Mexico Avenue
Howard Avenue	Newton
Milwaukee	Newton
Providence	Newton
Howard Avenue	Newton
Forsyth Avenue	Newton Place
Scott Avenue	Newton Place
Brads Avenue	Oak
Breeds Terrace	Oak
Laurel Avenue	Oak
Oak	Oak Court
O	Oates
Kansas	Ontario Place
Huron	Ontario Road
Ontario	Ontario Road
Ontario Avenue	Ontario Road

Original Street Names

Original Name	Became
Omaha	Ordway
Military Road	Oregon Avenue
Shepherd Road	Oregon Avenue
Fort	Otis
Lawrence	Otis
Newark	Otis
Cammack Avenue	Otis Place
Lamar Place	Otis Place
Lowell	Park Road
Park	Park Road
Whitney Avenue	Park Road
P	Penn
Monroe	Perry
Omaha	Perry
Cottrell Place	Potomac Avenue
R	Q
Madison	Quackenbos
Quincy	Quebec
Holbrook Terrace (north)	Queen
Q	Queen
Newark	Quincy
Philadelphia	Quincy
Quincy	Quincy Place
U	R
Omaha	Randolph
Perry	Randolph
Quincy	Randolph
R	Raum
Ridge	Reno Road
Randolph	Riggs Place
Nicholson	Rittenhouse
Cathedral Avenue	Rock Creek Drive
Otis	Rodman
Richmond	Rodman
Seaton	Rosedale
Linthicum Place	S
V	S
Seaton	Seaton Place
Savannah	Sedgwick
Philadelphia	Shepherd
Quincy	Shepherd
Oglethorpe	Sheridan

Original Street Names

Original Name	Became
Summit	Summit Place
California Avenue (w Columbia Rd)	T
Maple	T
W	T
Quincy	Taylor
Randolph	Taylor
Quincy	Tilden
Trenton	Tilden
Thomas	Todd Place
Irving	Tracy Place
Irving Place	Tracy Place
Peabody	Tuckerman
Madison	U
Vernon Avenue	U
Richmond	Upshur
Randolph	Upton
Utica	Upton
California Avenue (east of 19th)	V
Wilson	V
Y	V
Shepherd	Van Ness
Vallejo	Van Ness
Savannah	Varnum
Utica	Varnum
Tenley Place	Verplanck Place
Pomeroy	W
Walnut Avenue	Walnut
Warder Avenue	Warder
Xenia	Warren
Austin Place	Warren Place
Park Road	Waterside Drive
Trenton	Webster
Vallejo	Webster
Xenia	Webster
Boundary	Western
Susquehannah	Whittier
Willow Avenue	Willow
Austin	Windom Place
Woodley	Woodley Road

Orleans Place (NE)
See City Streets.

Otis Place (NW), Street (NE)
Described by John Adams as a "flame of fire," James Otis (1725-1783) fought for colonial rights by leading the American opposition to Britain. He, with Samuel Adams, protested the Townsend Acts of 1767, which taxed certain items imported to the colonies, such as tea. Their efforts helped to bring about the Boston Massacre and the Boston Tea Party.

Palisade Lane (NW)
Palisade Lane is located in the upper northwest corner of the District. The name is derived from a physical feature of the area just above Chain Bridge: a cliff carved out by the flow of the Potomac River.

Papermill Court (NW)
Paper Mill was located on P Street at Florida Avenue. The mill was built around 1800 by Gustavus Scott and Nicholas Lingan. The property was owned by Edgar Patterson until it was purchased by Elie Williams, Daniel Carroll of Duddington and Charles Carroll of Bellevue in 1811.

Park Road (NW)
Park Road was formerly a combination of Lowell Street, Whitney Avenue and Park Street. The Columbia Heights Citizens' Association fought for several years to have the streets renamed Park Avenue and, when that failed, Park Road. The Association argued that the name was perfect because the street was the only connection between the U.S. Soldiers Home and Rock Creek Park, "two great park systems in Washington." In August 1905, the commissioners stated that the streets would be renamed Marshall Road for several reasons: 1) The name Park Avenue would not conform to the plan—only a name

with two syllables beginning with M would satisfy the requirements; 2) the Senate Park Commission was planning a connection between the two parks farther north, and 3) the street was not wide enough at certain points to support the designation of avenue. In October 1905, the commissioners abandoned their plan to rename the street Marshall Road and officially adopted Park Road.

Whitney Avenue was named for Asa Whitney, who gave 9.86 acres to the government for the U.S. Soldiers Home. This entrance road to the Soldiers Home was named Tayloe's Road before the 1869 property transfer, which stipulated that the street bear the Whitney name. In 1873, two individuals, Stuart J. Gass and Catherine M. Whitney, gave money and property respectively for the widening of Whitney Avenue. In 1899, control of Whitney Avenue was transferred from the Soldiers Home to the commissioners.

Parkside Drive (NW)

Parkside is a descriptive name for this street. The street travels approximately three blocks along the eastern edge of Rock Creek Park as it approaches the Maryland border.

Patricia Roberts Harris Drive (NE)

Located between 31st Street and Fort Lincoln Drive, this street is named for lawyer Patricia Roberts Harris (1924–1988). Harris became the first black woman to serve in the cabinet when President Jimmy Carter appointed her Secretary of Housing and Urban Development in 1977. She later served as Secretary of Health and Human Services.

Patrick Circle (SW)

Marsena Rudolph Patrick (1811–1888) had strong moral values. Prior to the outbreak of the Civil War, Patrick served as the president of the New·York State Agricultural College. He despised corruption and he took a disciplinarian approach with his troops. He served as military governor of Fredericksburg and as provost marshal for all armies operating against Richmond. His duties included protecting civilians, guarding prisoners and deserters, regulating trade and keeping order among the troops. At the conclusion of the war, Patrick served as the provost at Richmond. Patrick's diary, *Inside Lincoln's Army*, was published in 1964.

Payne Terrace (SE)
William H. F. Payne (1830–1904) enlisted in the Confederate army when Virginia seceded from the Union. Payne was wounded and captured three times during the war. After the conflict, he served in the Virginia legislature and as general counsel for Southern Railway.

Pecan Street (SE)
See Flora and Fauna.

Penn Street (NE)
William Penn (1644–1718), the founder of Pennsylvania, had a special place in the hearts and minds of the residents of the new capital, since Congress moved to Washington from Philadelphia, a city he created. Penn created Pennsylvania as a refuge for Quakers. His relationship with the area Indians paved the way.

Pennsylvania Avenue (NW, SE)
Once referred to as the "Great Servonian Bog," Pennsylvania Avenue was envisioned by Pierre L'Enfant as the grandest avenue in the city, an honor it holds today. Until 1800, the street was not a grand width, though. It was widened from 35 feet to its present 160 feet, in keeping with L'Enfant's original design. The street, cut across a swampy thicket, travels the city from the Anacostia River to Rock Creek.

Since Pennsylvania Avenue is a direct route from the White House to the Capitol, it has been host to inaugural and military parades and ceremonies honoring foreign dignitaries. This status makes it first on any list for decorative and reparative work.

Individuals, especially presidents (possibly because they view the street more than anyone else), feel the need to constantly recommend improvements for the street. Thomas Jefferson wanted Pennsylvania Avenue to provide a pleasant, shady approach to the Capitol, so four rows of lombardy poplars were planted, one on each side and two down the middle. Jefferson believed that the poplars would be replaced by willow trees before they became too large. The replanting never occurred and the poplars overtook the pedestrian walkways. In 1833, at Jefferson's request, Congress ordered the trees cut down. During his presidency, John F. Kennedy had plans drawn to completely rebuild the avenue, but this massive reconstruction never materialized. (The traffic jam it would have caused is hard to imagine.)

Pennsylvania Avenue, in 1832, was one of the first streets to be paved, but only the 1.5 miles between the Capitol and the White House. Wood blocks were used first, and later, cobblestones. The street was badly damaged by the troops and equipment hauled over it during the Civil War and, in 1870, it was paved with asphalt as part of District of Columbia Governor Alexander Shepherd's controversial city redevelopment. (See Shepherd Street.)

Also during Shepherd's redevelopment, the buildings were upgraded to reflect the avenue's stature in the city. These buildings replaced the wooden ones erected years before. Federal Triangle was conceived in the 1920s and 1930s in an effort to return to non-war life after World War I. The project employed several thousand people.

When Congress said "Let there be light," Pennsylvania Avenue was again first. In 1842, the distance from the White House to the Capitol was lined with oil lamps. For seven years, it remained the only lighted street in Washington. Although electric lights were first demonstrated in 1879 at Thomas Circle, the first permanent ones were lit on this street (at 15th Street) in 1881.

Thomas Jefferson's inaugural parade was the first to travel down the avenue. Jimmy Carter was the first president to walk back down the avenue after his inauguration.

Pennsylvania Avenue has been the route for many presidential funeral processions, but in 1841 William Henry Harrison's was the first to plod from the White House to the Capitol.

It is difficult to imagine a time when cars did not travel down this street, but in 1846, the Duryea was the first. Public transportation exists today in the form of the subway and buses. Cable cars and streetcars ceased to exist in Washington when service was discontinued down Pennsylvania Avenue in the 1960s.

Pennsylvania Avenue's image has not always been pristine. In the 1890s, the new Post Office (100 years later, it is the Old Post Office) was erected. The building, described by one senator as "a cross between a cathedral and a cotton mill," was built in an effort to clean up the area. The seedy "Hooker's Division," named for the Civil War quarters of General Hooker's troops, ran between 10th and 15th Streets. A guide to Washington, *Night Side*, described the area in 1894: "If you are so lost to all self respect as to go there to witness drunkenness, ribald song, vulgarity, immodesty, and all the other vices, do not, as you value truth and honor, take any innocent youths with you. It has been the ruin of thousands of the bright and talented young men of this city. Watches, honor and happiness have been left there innumerable times. Only a fool complains of his losses. Wise men stay away."

Perry Place (NE, NW), Street (NE)

American naval hero Oliver Hazard Perry (1785-1819) won the decisive battle in the War of 1812. Perry is best known for this description of the 1813 Battle of Lake Erie: "We have met the enemy and they are ours."

Pershing Drive (NW)

Pershing Drive is named for General John Joseph Pershing (1860-1948). A veteran of the Spanish-American War, Pershing had an ability to quickly mold inexperienced men into a strong fighting force. Nicknamed Black Jack, he served as commander of the American Expeditionary Force during World War I (1917-18).

Phelps Place (NW)

Known for his superior debate techniques, John S. Phelps (1814-1886) served in the House of Representatives for 18 years. In 1862, he was appointed military governor of Missouri; 14 years later he was elected governor of the state.

Pierce Street (NW)

Pierce Street is named for the 14th president, Franklin Pierce (1804–1869). At the commencement of his political career, Pierce served in the New Hampshire legislature (1829–33) and in Congress (1833–42). He decided to return to his home state to practice law, resigning his congressional seat. Ten years later, Pierce was nominated as a dark horse candidate of the Democratic Party. The party at this time was divided and Pierce's nomination was a compromise. Despite this obstacle, he soundly defeated Whig candidate Winfield Scott. In 1853, Pierce effected the Gadsden Purchase, the acquisition of a strip of land near the Rio Grande River. This purchase from Mexico allowed for the advancement of a southern railroad route. But this victory was overshadowed by Pierce's support of the Kansas-Nebraska Act of 1850, which enraged Northerners. The Democrats wrote him off; he was not considered for nomination in 1856. Pierce was vocal about his opposition to the Civil War, a stance that made him even more unpopular in the North.

Pierce Mill Road (NW)

Pennsylvanian Isaac Peirce (1756–1841) owned 2,000 acres from the present day National Zoological Park to Chevy Chase. Along his Rock Creek property, he and his son, Abner C. Peirce, built a grist mill, a saw mill, a house and barns. In 1890, Peirce (often spelled Pierce) sold 350 acres to the U.S. government for the sum of $250,000 to aid in the creation of Rock Creek Park. The mill, which began production in 1820, was last operated in 1897. The street name and the mill are all that remain of Peirce's property.

Piney Branch Parkway (NW), Road (NW)

These two streets take their names from the Piney Branch of Rock Creek.

Polk Street (NE)
Polk Street is named for the 11th president of the United States, James K. Polk (1795-1849). Polk's political career began with service in the Tennessee legislature (1823-25) and Congress (1825-39). Polk served as speaker of the House from 1835 to 1839. In 1839, Polk became governor of Tennessee, but he failed to be re-elected. In 1844, the Democrats could not decide upon a candidate. Polk was presented as a compromise. He defeated Henry Clay for the presidency by a narrow margin. Once elected, Polk kept his campaign promises: reduction of tariff, settlement of the Oregon boundary dispute, annexation of California and redevelopment of an independent treasury. He sent troops to the Rio Grande and caused the Mexican War, which resulted in the acquisition of the Southwest.

Pond Street (NE)
Pond Street is located on the eastern side of the Kenilworth Aquatic Gardens, an area that contains several ponds.

Pope Street (SE)
Pope Street is named for the designer of the Jefferson Memorial, John Russell Pope. Pope made several architectural marks on the city, including the National Gallery of Art and the National Archives.

Poplar Street (NW, SE)
See Flora and Fauna.

Porter Place (NW), Street (NE, NW)
Porter Street is named for the Porter family of naval heroes. David Porter (1780-1843) was a hero of the War of 1812. His ship, the *Essex*, won several battles against the British. His sons, David Porter (1813-1891) and William D. Porter (1809-1864), fought during the Civil War. David took part in the seige on New Orleans and

the capture of Arkansas Post and Vicksburg. At the end of the war, he served as commander of the North Atlantic Blockading Squadron. William, whose ship was named after his father's, saw action on the Tennessee River and on the Mississippi, participating at Natchez, Vicksburg and Port Hudson. William died of heart disease before the end of the war. William's two sons fought the war for the Confederacy. David Porter's adoptive son was David Farragut (see Farragut Street).

Portland Street (SE)
See City Streets.

Potomac Avenue (SE, SW)
Potomac Avenue is in two sections: One is located near Fort Lesley J. McNair, the other runs between the Navy Yard and Congressional Cemetery. Although named for the Potomac River, both streets are situated closer to the Anacostia River.

Powhatan Place (NW)
Indian leader Wahunsonacock (c. 1550–1618) headed the Powhatan Confederacy. The chieftain was referred to as Powhatan (the English translation means a fall or rapid in a stream) as an official title. The legend of Pocahantas says that Powhatan, opposing settlers along the James River, was about to kill Captain John Smith when his daughter Pocahontas saved him. Nevertheless, Powhatan maintained a tense peace between the Indians and the settlers. After his death, war broke out between these two peoples.

Presidents
The following District streets are named for U.S. Presidents.

Adams Street	John Adams (1797–1801)
	John Quincy Adams (1825–1829)
Buchanan Street	James Buchanan (1857–1861)

Cleveland Avenue	Grover Cleveland (1885–1889/ 1893–1897)
Eisenhower Freeway	Dwight D. Eisenhower (1953–1961)
Garfield Street	James Garfield (1881)
Grant Street	Ulysses S. Grant (1869–1877)
Harrison Street	William Henry Harrison (1841) Benjamin Harrison (1889–1893)
Hayes Street	Rutherford B. Hayes (1877–1881)
Jackson Street	Andrew Jackson (1829–1837)
Jefferson Street	Thomas Jefferson (1801–1809)
Johnson Avenue	Andrew Johnson (1865–1869)
Kennedy Street	John F. Kennedy (1961–1963)
Lincoln Road	Abraham Lincoln (1861–1865)
Madison Street	James Madison (1809–1817)
McKinley Street	William McKinley (1897–1901)
Monroe Street	James Monroe (1817–1825)
Pierce Street	Franklin Pierce (1853–1857)
Polk Street	James K. Polk (1845–1849)
Taylor Street	Zachary Taylor (1849–1850)
Van Buren Street	Martin Van Buren (1837–1841)
Washington Circle	George Washington (1789–1797)

It is easy to understand why streets are not named for living presidents (since a law prohibits naming streets for the living), but several presidents were forgotten in the naming process. The following have not been recognized:

John Tyler (1841–1845)
Millard Fillmore (1850–1853)
Chester A. Arthur (1881–1885)
Theodore Roosevelt (1901–1909)
William Howard Taft (1909–1913)
Woodrow Wilson (1913–1921)
Warren G. Harding (1921–1923) (died in office)
Calvin Coolidge (1923–1929)
Herbert Hoover (1929–1933)
Franklin D. Roosevelt (1933–1945) (died in office)
Harry S Truman (1945–1953)
Lyndon Johnson (1963–1969)

Primrose Road (NW)
See Flora and Fauna.

Princeton Place (NW)
See City Streets.

Prospect Street (NW)
Prospect Street formed the northern border of the city of Georgetown, Maryland. It also served as the property line and derived its name from Benjamin Stoddert and Uriah Forrest's country home, Pretty Prospect. Legend has it that Stoddert used to watch for his ships to return to port through a telescope situated at the southern windows of his Halcyon House. Seventy-two acres of the property, later owned by Mrs. M. C. Barber, were sold to the government for the construction of the U.S. Naval Observatory. (See Massachusetts Avenue/Observatory Circle.)

Providence Street (NE)
See City Streets.

Puerto Rico Avenue (NE)
Puerto Rico Avenue is named for the American territory. Puerto Rico is a self-governing body. Citizens have American citizenship, but they do not pay federal income taxes or vote in U.S. elections.

Quarles Street (NE)
William A. Quarles (1825–1893) was a circuit court judge and a railroad president until the outbreak of the Civil War. He was elected to lead a Tennessee infantry regiment. He fought bravely in several battles, but was wounded and captured twice during the war. After his second release, he returned to his law practice in Tennessee. He served in the Tennessee senate and as a delegate to two Democratic conventions, where he wielded much power in the party.

Quarry Place (NW)
In the late 1700s, before the city of Washington existed, a rock quarry was located on the banks of Rock Creek.

Quarry Place is located near that spot. It is believed this is how the street derives its name.

Quebec Place (NW), Street (NW)
See City Streets.

Queen Anne's Lane (NW)
This street is named for Queen Anne of England (1665–1714). Her name was also taken for Queen Anne's County, Maryland.

Quincy Street (NE, NW)
Quincy Street is named for patriot Josiah Quincy. In 1905, the name was changed to Quincy Place to comply with the plan of the Commissioners of the District of Columbia. It was later returned to its current status.

Railroad Avenue (SE)
Aptly named, this block-long street travels along the railroad tracks in Anacostia.

Raleigh Street (SE)
Sir Walter Raleigh (c. 1554–1618) organized many journeys to America for potential colonization. Raleigh created a colony on Roanoke Island, Virginia, known as the "lost colony" because the second group of colonists to arrive disappeared without a trace. In 1595, Raleigh travelled to the Orinoco River, but with the reign of James I, he was convicted of treason and imprisoned upon his return to England. He was released in 1616. In search of gold, he promptly journeyed to Orinoco River again. He found nothing. He was executed for treason when he returned to England.

Randolph Street (NE, NW)
Edmund Randolph (1753–1813) exerted much influence on the Constitution by proposing the Virginia plan, which favored larger states. Randolph served as Amer-

ica's first attorney general (1789–93) and its second secretary of state (1794–95)

Randolph Place (NW)
Randolph Place is actually a continuation of Riggs Place. Residents of this two-block stretch west of North Capitol Street were infuriated by the change from Randolph Street that they petitioned the commissioners, asking to replace the "Street" designation with "Place." The commissioners obliged on May 17, 1905.

Raoul Wallenberg Place (SW)
Raoul Wallenberg (1912–1947?), a Swedish diplomat stationed in Budapest, saved approximately 20,000 Hungarian Jews from Nazi death camps by issuing them Swedish passports. When the Soviet army reached Budapest, Wallenberg was arrested. His death was reported by the Soviets in 1947, although it was never confirmed. Wallenberg was made an honorary citizen of the United States in 1981. A section of 15th Street, between Maine and Independence Avenues, was named in his honor in 1983.

Redbud Lane (NW)
See Flora and Fauna.

Redwood Drive (SE), Terrace (NW)
See Flora and Fauna.

Reno Road (NW)
Reno Road takes its name from nearby Fort Reno. The fort, one of many built in the defense of Washington during the Civil War, is named for Jesse Lee Reno (1823–1862).

Reno, a veteran of the Mexican War, commanded arsenals in Alabama and Kansas until the Civil War broke out. At the outset, Reno commanded a brigade under Brigadier General Ambrose E. Burnside. He later fought at the Second Battle of Bull Run and assisted in the pursuit of Stonewall Jackson's cavalry through Maryland. Reno, a well-respected commander, was killed at the Battle of South Mountain while running up and down the lines offering words of encouragement. A story of the time states that Reno offered to buy a flag from a woman in Frederick, Maryland. She had waved the flag, it was thought, as she watched Jackson march through town. She, in fact, had not seen Jackson, but she gave Reno a homemade flag. He carried the flag in his saddlebag. When he was buried, the flag was draped over his casket. Shortly after, Fort Pennsylvania was renamed in his honor.

During the 1950s, attempts were made to change Reno Road to Washington Avenue. These efforts failed. Washington Avenue was created in 1989. (See Washington Avenue.)

Reservoir Road, Street (NW)

Reservoir Road stretches from the Georgetown Reservoir, located adjacent to the C&O Canal, to the Georgetown branch of the D.C. public library, the location of the old Georgetown Reservoir. The wall at the property line on Wisconsin Avenue remains from the old reser-

RESERVOIR ROAD. This street runs from the original Washington reservoir at Wisconsin Avenue to the present city reservoir at MacArthur Boulevard. The Georgetown Branch of the D.C. Public Library is now located on the original site. The stone wall around the property remains from the original reservoir.

voir. Water is still conducted through the original pipes. (See MacArthur Boulevard.)

Rhode Island Avenue (NW)
See State Streets.

Rhododendron Valley Road (NE)
See Flora and Fauna.

Richardson Place (NW)
This street is named for William A. Richardson, who served as secretary of state (1873–74) under Ulysses S. Grant.

Riggs Road (NE)
Riggs Road is named for the Riggs Mill once located on the street. Formerly known as the Adelphi Mills, the property had four different owners until the Riggs family, headed by George Washington Riggs, took it over and changed the name in 1860.

Riggs Place (NW)
When residents complained about the renaming of Randolph Street, Major John Biddle, U.S. Corps of Engineers, responded, "it is an old Washington name and it was thought suitable to give it to this street . . . ," but conceded and allowed the Randolph name to remain. (See Randolph Place.)

The street is named after George Washington Riggs, Jr., co-founder of the Corcoran-Riggs Bank. The bank, founded in 1840, later became Riggs Bank.

Rittenhouse Street (NE, NW)
Rittenhouse Street is named for David Rittenhouse (1732–1796), who is considered America's first astronomer. With his homemade instruments, Rittenhouse determined the boundaries of several states and a section

of the Mason-Dixon line. He also served as the first director of the U.S. Mint (1792–95).

River Road (NW)
Although it is difficult to tell from its route through the District of Columbia, River Road is aptly named. The street travels along the Potomac River in Montgomery County, Maryland.

Rock Creek and Potomac Parkway (NW)
This thoroughfare runs from the National Zoo and around toward the monuments near the Potomac River. The parkway, a continuation of Beach Drive, was constructed in the late 1920s to reduce traffic and provide easier access to downtown Washington.

Rock Creek Church Road (NW)
Rock Creek Church stood on this road. According to church records, it was built in 1775 and remodeled 93 years later. The church was destroyed by fire in 1921.

Rock Creek Ford Road (NW)
Before the area around Rock Creek was developed, the only way to cross the creek was at fords, shallow and narrow sections of the creek. On their journey north, Generals Rochambeau and Washington forded the creek farther down. As the area became more populated, bridges replaced the fords. This road is all that remains of one of these early crossing points.

Rodman Street (NW)
Thomas J. Rodman (1815–1871) patented a method of gun casting that made the cannon tubes stronger and better able to withstand heavy use. His method cooled the tubes from the inside, circulating water through the core. The Columbiad was designed by Rodman and utilized in many seacoast fortifications during the Civil

War. Rodman also developed gunpowder for use in heavy weapons.

Rosedale Street (NE)
Rosedale was the country home of Uriah Forrest, a Revolutionary War general and a member of the Continental Congress. The home was located near 34th and Newark Streets, NW.

Rosemount Avenue (NW)
Rosemount Avenue is the remnant of the country home of Edward Johnston, a long time resident of the area. Although Johnston's name does not remain, the name of his property did.

Ross Drive (NW), Place (NW)
J. W. Ross oversaw the construction of much of Rock Creek Park during the late 1800s. Ross Drive, which crosses a section of the Park above the National Zoo, was constructed and named while Ross was in command.

S

St. Louis Street (SE)
See City Streets.

Salem Lane (NW)
See City Streets.

Saratoga Avenue (NE)
Saratoga Avenue is named for the Battles of Saratoga during the Revolutionary War. These battles were fought between June and October 1777. The Americans defeated the British for the first time in the war; it is considered a turning point of the war.

Savannah Place (SE), Street (SE), Terrace (SE)
See City Streets.

Scott Circle (SW), Place (NW), Circle Drive (NW)

Lt. General Winfield Scott (1786–1866) was a hero of the War of 1812 and supreme commander of the U.S. Army during the Mexican War. At the outbreak of the Civil War he refused a Confederate commission, instead choosing to direct the defenses of Washington. Nicknamed "Old Fuss and Feathers" because of his adherence to proper military dress and protocol, Scott was defeated at Ball's Bluff in 1861. He resigned his command amid speculation about his senility. A superior strategist, his Anaconda Plan, a divide and conquer method of defeating enemy troops, blockaded the coast from southern Virginia to southern Texas and utilized the border states as a buffer zone.

A statue on the circle shows Scott atop a stallion. Although Scott always rode a mare into battle, his relatives felt it would be more fitting for the general to be seen riding a male horse. The sex of the horse was changed just before the statue was cast.

Seaton Place (NE, NW)

William Seaton, brother-in-law of Joseph Gales, served as co-owner and co-editor of the *National Intelligencer*. For a long time, the *Intelligencer* was the only newpaper published in Washington. The paper enjoyed strong government support; it provided a voice for three administrations: Adams, Jefferson and Madison. The newspaper folded around the start of the Civil War. Seaton served in political office as mayor of Washington from 1840–1850. Formerly Seaton Street, the name was changed in 1905. (See Gales Street.)

Sedgwick Street (NW)

Major General John Sedgwick (1813–1864) had a reputation as a dependable commander. He fought with distinction at Antietam, where he was badly wounded, and at Gettysburg, after a forced march delivered his troops to the battlefield in time. He was mortally wounded prior to the Battle of Spotsylvania, Virginia.

Service Road (NE)
Service Road is located on the grounds of the National Arboretum.

Shepherd Road (NW), Street (NE, NW)
After the Civil War, Washington was a mess. The streets had huge ruts from the weight of the equipment and troops, the Washington Canal was a polluted mosquito breeding ground and the existing parks had been neglected. In October 1869, representatives of Western and Southern states met in St. Louis and agreed to move the capital to a better location.

The citizens of the capital city met in late 1870 to discuss a single District government. The city was presently governed by three bodies: the city of Georgetown, Washington City and the Levy Court, which oversaw all land outside the two towns. In February 1871, Ulysses S. Grant signed the bill that created a territorial form of government for the District. Hoping to bolster the morale of the city's residents and prevent the moving of the capital, the government promised to initiate mass improvement efforts. As Vice-Chairman of the Board of Public Works, Alexander Robey Shepherd (1835–1902) supervised much of the work "in order that . . . the capital of the nation might not remain a quarter of a century behind the times." The city improvements included installing gas lights, digging sewers, grading and paving 80 miles of streets and planting 50,000 trees. Shepherd was appointed governor of the District in 1873. He continued the improvements, initiating several building projects: the National Museum (now the Smithsonian), completed in 1881; the Washington Monument, completed in 1884; the Pension Building, opened in 1885; the State, War and Navy Building (Old Executive Office Building), completed in 1888, and the Library of Congress, opened in 1897. The State, War and Navy Building was, at that time, the world's largest office building.

Shepherd's heart may or may not have been in the right place, but his love of Washington was intense. A shrewd businessman, Shepherd purchased J. W. Thompson, a plumbing and gas fitting company. During the

Civil War, he amassed a fortune selling supplies to the government. After the war, Shepherd, in addition to his responsibilities as an elected official, became involved in real estate, becoming the first man to build blocks of rowhouses in Washington. Shepherd also knew how to get what he wanted. During the redevelopment, Shepherd expressed a desire to have temporary railroad tracks removed from Pennsylvania Avenue near the Capitol. The railroad company refused to comply, instead placing a locomotive on the crossing. A few nights later, Shepherd ordered his men to remove the tracks, leaving only those directly under the engine.

The redevelopment ran the city government close to bankruptcy; in three years, Shepherd spent $20 million. The debt, which was greater than that incurred by both Washington and Georgetown in 70 years, forced Congress to reassume control of the city government in 1878. Shepherd was accused of corruption. An investigation launched in 1872 found no evidence of wrongdoing. Another investigation was ordered in 1874. As the city debt rose, Shepherd and his close friends enjoyed the finest things money could buy. One resident voiced the sentiments of many: "If anybody knows all about sewers, it must be Boss Shepherd." No one could ever prove Shepherd guilty of the charges, but his office and the government were abolished in 1874. President Grant appointed him head of the new government, but Congress tabled the request. Shepherd, forced out of the city by the controversy, took refuge in Mexico.

In 1887, though, Shepherd returned from Mexico to a huge celebration in his honor. Citizens carried signs that said: "Washington suggested, Congress sanctioned, Shepherd made it." At about the same time, Washington residents pledged money to erect a statue of Shepherd in front of the District Building. The first native Washingtonian to be so honored, Shepherd is appropriately depicted with one hand holding a map and the other behind his back.

Regardless of his guilt or innocence, the fact still remains that Washington was once again resurrected from the swamp and transformed into a model federal city. The only significant drawback was that street grading stranded many houses, especially in Georgetown, an effect that is still evident today.

Sheridan Circle (NW), Drive (NW), Road (SE), Street (NE, NW)

Sheridan Circle was named and a statue erected in honor of General Philip H. Sheridan (1831–1888), cavalry commander of the Union Army of the Potomac during the Civil War. Sheridan's "scorched earth" policy of burning Southern farms to cut Confederate food supplies was frowned upon by many. Nevertheless, he earned a reputation for effectively cutting Confederate supply and transportation lines. After the war, Sheridan was appointed commander-in-chief of the armed forces.

The statue on the circle, located at Massachusetts Avenue and 23rd Street, features Sheridan astride his favorite horse Rienzi. Sheridan renamed the horse Winchester after rallying his troops to a victory at the Battle of Cedar Creek, near the Virginia city of Winchester. The statue, designed by Gutzon Borglum (of Mt. Rushmore fame) was erected in 1909 and dedicated by President Theodore Roosevelt.

Sherman Avenue (NW)

Sherman Avenue is named for John Sherman (1823–1900), brother of Civil War General William Tecumseh Sherman (see Sherman Circle). An extremely successful trial lawyer turned politician, Sherman is best known for lending his name to the Sherman Anti-Trust Act. He was spirited into politics by his strong opposition to the Kansas-Nebraska Act. Sherman was elected to the House of Representatives as a Whig and returned for three more terms as a Republican. He served 32 years in the Senate and made three unsuccessful bids for the presidency. His reputation as a financial and military expert brought him appointments as Rutherford B. Hayes's Secretary of the Treasury and William McKinley's Secretary of State.

Sherman Circle (NW)

Sherman Circle, located at Kansas and Illinois Avenues, is named for Civil War general William Tecumseh Sherman (1820–1891). Famous for his fiery march to the sea after capturing Atlanta, Sherman was criticized by

Southerners for his destruction of property. He contended that wasting property was better than wasting lives in battle. Sherman, brother of politician John Sherman, was appointed general-in-chief of the army when Ulysses S. Grant was inaugurated. (See Sherman Street.)

Sixteenth Street

Running six and a half miles from the Maryland border at Silver Spring to Lafayette Square, Sixteenth Street was intended to serve as a ceremonial approach to Washington. It is the second longest street in the District of Columbia (Massachusetts is first).

The socialite residents of the Meridian Hill area, at the prompting of Maryland Senator John B. Henderson's wife, requested that the White House be moved from Pennsylvania Avenue to their neighborhood. (Senator Henderson of Missouri drafted the Constitutional amendment abolishing slavery.) The Hendersons purchased substantial amounts of land, built elaborate houses and hoped that the embassies would flock to the area and form Embassy Row on 16th Street. The only accomplishment achieved by Mrs. Henderson's efforts was changing the street name to Executive Avenue (from H Street to the Maryland border) in 1906 and to Avenue of the Presidents in 1913 (March 4, 1913, to July 24, 1914, to be exact).

Sixty-Third Street

Located in the southeast corner of the District, 63rd Street is the highest-numbered street in the city.

South Capitol Street (SE, SW)

South Capitol Street serves as the dividing line between the southeast and southwest quadrants of the District of Columbia.

South Carolina Avenue (SE)

See State Streets.

South Dakota Avenue (NE)
See State Streets.

Southern Avenue (SE)
Southern Avenue marks the southern border of Washington. Separated into three parts as it travels the Prince George's County line, the street runs from South Capitol Street to Naylor Road, Branch Avenue to Benning Road and Central Avenue to Eastern Avenue.

South Portal Drive (NW)
South Portal Drive forms the southern boundary to the Montgomery Blair Portal, an entrance into Rock Creek Park.

Southeast-Southwest Freeway
Renamed Dwight D. Eisenhower Freeway in 1991, this highway was constructed in 1967. Eisenhower, who served as president from 1953 to 1961, played a large role in the initiation of the interstate highway system and affected the redevelopment of southwest Washington.

Spring of Freedom Street (NW)
A one-block section of Linnean Street, between Tilden and Shoemaker Streets, was renamed in 1991 to honor the 1989 dismissal of communist governments in Eastern Europe. The occupants of that block are the Hungarian and Czechoslovakian ambassadors.

Spruce Drive (NW), Street (SE)
See Flora and Fauna.

Stanton Road (SE), Terrace (SE)
Edwin M. Stanton (1814–1869) served as Abraham Lincoln's secretary of war. Stanton's unyielding desire to keep the Union together caused many arguments be-

tween him and the president. Stanton held the cabinet position into Andrew Johnson's presidency. In 1868, he resigned his office, a move that prompted impeachment proceedings against Johnson, and refused to run for any political office. In 1869, Ulysses S. Grant nominated him for a Supreme Court seat, but the confirmation came four days after Stanton's death.

State Streets

According to the original plan for the city, all diagonal streets would be named for states in the Union. All 50 states were not represented until 1989. The plan called for the streets to be placed in the same general geographical location as is seen on a map. Most of the streets are positioned correctly; some, named later, do not satisfy the requirement. The following is a list of the states represented, the date they entered the union and the number:

State	Year (No.)	Street
Alabama	1819 (22)	Alabama Avenue
Alaska	1959 (49)	Alaska Avenue
Arizona	1912 (48)	Arizona Avenue
Arkansas	1836 (25)	Arkansas Avenue
California	1850 (31)	California Street
Colorado	1876 (38)	Colorado Avenue
Connecticut	1788 (5)	Connecticut Avenue
Delaware	1787 (1)	Delaware Avenue
Florida	1845 (27)	Florida Avenue
Georgia	1788 (4)	Georgia Avenue
Hawaii	1959 (50)	Hawaii Avenue
Idaho	1890 (43)	Idaho Avenue
Illinois	1818 (21)	Illinois Avenue
Indiana	1816 (19)	Indiana Avenue
Iowa	1846 (29)	Iowa Avenue
Kansas	1861 (34)	Kansas Avenue
Kentucky	1792 (15)	Kentucky Avenue
Louisiana	1812 (18)	Louisiana Avenue
Maine	1820 (23)	Maine Avenue
Maryland	1788 (7)	Maryland Avenue
Massachusetts	1788 (6)	Massachusetts Avenue
Michigan	1837 (26)	Michigan Avenue
Minnesota	1858 (32)	Minnesota Avenue
Mississippi	1817 (20)	Mississippi Avenue

State Streets

Missouri	1821 (24)	Missouri Avenue
Montana	1889 (41)	Montana Avenue
Nebraska	1867 (37)	Nebraska Avenue
Nevada	1864 (36)	Nevada Avenue
New Hampshire	1788 (9)	New Hampshire Avenue
New Jersey	1787 (3)	New Jersey Avenue
New Mexico	1912 (47)	New Mexico Avenue
New York	1788 (11)	New York Avenue
North Carolina	1789 (12)	North Carolina Avenue
North Dakota	1889 (39)	North Dakota Avenue
Ohio	1803 (17)	Ohio Drive
Oklahoma	1907 (46)	Oklahoma Avenue
Oregon	1859 (33)	Oregon Avenue
Pennsylvania	1787 (2)	Pennsylvania Avenue
Rhode Island	1790 (13)	Rhode Island Avenue
South Carolina	1788 (8)	South Carolina Avenue
South Dakota	1889 (40)	South Dakota Avenue
Tennessee	1796 (16)	Tennessee Avenue
Texas	1845 (28)	Texas Avenue
Utah	1896 (45)	Utah Avenue
Vermont	1791 (14)	Vermont Avenue
Virginia	1788 (10)	Virginia Avenue
Washington	1889 (42)	Washington Avenue
West Virginia	1863 (35)	West Virginia Avenue
Wisconsin	1848 (30)	Wisconsin Avenue
Wyoming	1890 (44)	Wyoming Avenue

Stevens Road (SE)

Thaddeus Stevens (1792–1868) led the Radical Republicans' program of Civil War Reconstruction. After the war, Stevens considered the defeated Confederate states "conquered provinces." While serving in the House of Representatives, Stevens suggested the 14th Amendment (civil rights). Stevens also supported the impeachment proceedings against Andrew Johnson.

Stoddert Place (SE)

Stoddert Place is named to honor Benjamin Stoddert, an important figure in the commercial and political life of Georgetown. Stoddert also helped to create the Bank of Columbia, which handled extensive real estate transfers during the development of the federal city. Stoddert and William Deakins, Jr., privately purchased land, at George Washington's request, in an attempt to avoid the speculation that would drive the real estate costs up. Stoddert went on to serve as Secretary of the Navy under John Adams.

Stuyvesant Place (NW)
Peter Stuyvesant (1610–1672) was the Dutch director-general of the North American colony, New Netherland. He expanded the colony by attacking and conquering New Sweden (today's New Castle, Delaware). Eight years later, the one-legged Stuyvesant battled with the British over the same territory . . . briefly. He surrendered to the British in 1664.

Sudbury Lane (NW), Place (NW), Road (NW)
See City Streets.

Suitland Parkway (SE)
Suitland Parkway runs from South Capitol Street through, you guessed it, Suitland, Maryland. The town itself was named for Colonel Samuel Taylor Suit, a state senator who resided in the area.

Sumner Road (SE)
Charles Sumner (1811–1874) was a dedicated senator whose outspokenness often inflamed his Congressional colleagues. After giving a speech opposing the Kansas-Nebraska Act, Sumner was beaten unconscious by Senator Preston Brooks of South Carolina. As the leading Radical Republican, Sumner motivated the impeachment proceedings against Andrew Johnson. Sumner died of a heart attack in the Senate chamber; he lay in state in the Capitol rotunda.

Sycamore Drive (SE), Place (NE), Street (NW)
See Flora and Fauna.

Tamarach Street (NW)
See Flora and Fauna.

Taylor Street (NE, NW)
Taylor Street is named for "Old Rough and Ready" Zachary Taylor (1785–1850), 12th president of the U.S. Taylor earned his nickname in the Black Hawk War and in battles against the Seminole Indians in Florida. Taylor's greatest popularity came during the Mexican War; while commanding an army in Texas, he won many decisive battles. Taylor was elected president in 1848. Charges of corruption were made against his cabinet, but he died of cholera before an investigation could be launched. He was succeeded by Millard Fillmore.

Tenley Circle (NW)
The Tenleytown neighborhood was created from land owned by Colonel Thomas Addison and James Stoddert. Whom Tenley Circle and the surrounding neighborhood are named for depends upon whom you ask. One story says it was named for a Maryland blacksmith who worked at the present-day intersection of River Road and Wisconsin Avenue. Another story says Tenley was the name of two sibling weavers who lived near the toll gate. Yet another tale states the name is to honor John Tennally, a pub owner. The Tennally option seems to have the most substantiation in the records.

Tennally owned a tavern near the intersection of Wisconsin Avenue and River Road. It was the first food and beverage sales location built above Georgetown.

A tollbooth was constructed about 1829 near Tennally's Tavern. The toll was established to raise money for improvements to Frederick Pike (Wisconsin Avenue) and the National Road (U.S. 40).

By the early 1900s many different spellings of Tennally were being used. In 1920, to put an end to the confusion, the Post Office declared Tenleytown the official spelling. Tenley Circle derives its name from the neighborhood.

Tennessee Avenue (NE)
See State Streets.

Tennyson Street (NW)
Tennyson Street is named for poet Alfred Lord Tennyson (1809–1892). Tennyson's works include "Morte d'Arthur," "Ulysses" and "Locksley Hall." Tennyson supported Victorian values, and became that age's most famous poet.

Texas Avenue (SE)
See State Streets.

Thayer Street (NE)
Major General John M. Thayer (1820–1906) was known not for his military performance, but for his political career after the Civil War. Thayer served as one of the first senators from Nebraska. When his reelection bid failed in 1871, Ulysses S. Grant appointed him governor of the Wyoming Territory. He later served six years as the governor of Nebraska.

Thomas Circle (NW), Street (NW)
Thomas Circle, which honors General George H. Thomas (1816–1870), joins Massachusetts and Vermont Avenues. A Virginia native, Thomas alienated his friends and family to fight for the Union. The U.S. government, though, questioned his intentions. He fought valiantly for the Union cause, earning the nickname "Rock of Chickamauga" for holding his line against Confederate attack until support arrived. Thomas served as Sherman's second in command during the Atlanta Campaign, but the government still distrusted him. He finally earned Union respect after two days of intense fighting to keep Nashville, Tennessee, in Union control.

A statue of Thomas was built in 1870 to celebrate the end of the war. A two-hour military parade and the first demonstration of electric lights marked the unveiling of the statue on November 19, 1879. The Portland, Washington's first apartment house, was located on Thomas Circle.

Thornton Place (NW)

As with the President's House, the commissioners of the capital city held a competition to determine the best design for the Capitol. Dr. William Thornton (1759–1828) submitted the winning design in 1793. Thornton, trained as a physician, wanted to please Thomas Jefferson and kept altering the original to satisfy Jefferson's preferences, since Jefferson would influence the final decision. The Capitol by 1848, though, had become too small for the number of state representatives and had to be enlarged. The expansion was completed in 1863. Thornton, who said he often regretted not studying architecture, went on to design several buildings in Washington, including St. John's in Georgetown and Octagon House on Lafayette Square.

He also became extremely active in District politics, serving as commissioner of the Federal Buildings (1794), first superintendent of the Patent Office and vice president of the D.C. Medical Society. In spite of all his political accomplishments, August 25, 1814, is the day Thornton's name became cemented in Washington history. As the British were burning the city of Washington, Thornton, it is said, saved the Patent Office from certain incineration. Thornton stood on the steps and told the British officer, "To burn what would be useful to all mankind, would be as barbarous as formerly to burn the Alexandrian Library, for which the Turks have since been condemned by all enlightened nations."

Thornton, a lover of horses, established Washington's first race course in 1801. President John Quincy Adams was said to walk from the White House to enjoy the races.

Tilden Place (NW), Street (NW)

A lawyer, Samuel Jones Tilden (1814–1886) became drawn to politics. In 1874, he was elected governor of New York. Two years later, the Democrats nominated him for president. His Republican opponent was Rutherford B. Hayes. Tilden won the popular vote, but the electoral votes of several states were disputed. A congressional council (consisting of eight Republicans and seven Democrats) was formed to decide the election. Hayes emerged victorious by a one-electoral-vote margin.

Todd Place (NE)

In 1905, residents suggested Todd Place be changed back to the original name, Thomas Street. The commissioners denied the request, stating that Todd Place had already been placed in the official surveyor records. They also cited the existence of Thomas Circle, noting that duplication of names was to be avoided.

Tracy Place (NW)

Benjamin F. Tracy served as Secretary of the Navy (1889–93) under Benjamin Harrison.

Trenton Place (SE), Street (SE)

See City Streets.

Trinidad Avenue (NE)

Trinidad Avenue is named for the West Indian island, sister to Tobago. The two islands, discovered by Christopher Columbus in 1498, became politically joined in 1888. They became an independent state in 1962.

Tuckerman Street (NE, NW)

Tuckerman Street is named for Walter Tuckerman. Tuckerman's family were original residents of the capital city.

Tudor Place

This street was once owned by Thomas and Martha Peter. The couple used an $8000 bequest from Martha's grandfather, George Washington, to purchase the block.

Tulip Street (NW)

See Flora and Fauna.

Tunlaw Road (NW)

Adolphus Pickrell purchased the land that became the Tunlaw Farm in 1862. The property name was suggested by Ulysses S. Grant. The name memorializes the farm's most prominent tree, a magnificent walnut that stood out front. (If you haven't noticed yet, Tunlaw is walnut spelled backwards.)

The farm was situated between Foxhall Road and Wisconsin Avenue. Tunlaw Road, which today is a continuation of New Mexico Avenue, was the main road through the land. The street name first appeared in directories in the 1880s.

Another theory about the naming of Tunlaw Road stems from the Civil War. Some state that the name is derived from a code name for a Union route. This route is said to have passed through a walnut grove.

Underwood Place (NW), Street (NW)

Joseph R. Underwood was a senator and judge.

Universities

Six streets in the District of Columbia were named for American colleges. The streets originally existed side by side in the Mount Pleasant area. See the individual street listings for information on the surviving streets.

Columbia Road is still in existence.
Dartmouth Street was renamed Lamont Street in 1905.
Harvard Street still exists.
Kenyon Street still exists.
Princeton Street ceased to exist, but it is survived by Princeton Place.
Yale Street became Fairmont Street in 1905.

University Avenue (NW), Place (NW), Terrace (NW)

University Avenue travels along the western border of the American University campus. University Terrace is located a block away.

University Place is in the Mount Pleasant neighborhood. Although no college is situated nearby, it is close to streets named for institutions of higher learning (i.e., Columbia Road, Harvard Street, etc.).

Upshur Street (NE, NW)
Abel P. Upshur served as John Tyler's Secretary of the Navy (1841–1843) and Secretary of State (1843–1844). Upshur was rarely in the spotlight . . . until his death. Upshur, while attending a weapons demonstration aboard the USS *Princeton*, was killed when the 27,000-pound "Peace Maker" cannon exploded. The river disaster on February 28, 1844, was considered one of the city's worst.

Upton Street (NW)
A strong advocate of military reform, Emory Upton (1839–1881) proved himself to be an outstanding commander in all three branches of the army: infantry, cavalry and artillery. It was with the latter that he fired the first shots at the first Battle of Bull Run. Upton was a strong believer that the U.S. military system needed reform. A student of the German military, Upton developed attack techniques that are still in use today. In an effort to end his violent headaches, Upton committed suicide at the Presidio in San Francisco. A National Defense University building on the grounds of Fort Lesley J. McNair is named in his honor.

Utah Avenue (NW)
See State Streets.

Van Buren Street (NW)
Martin Van Buren (1782–1862) served as the eighth president of the United States. Van Buren was very active in Democratic politics. He served in the Senate (1821–28) and was elected governor of New York in 1828. He resigned one year later to accept an appointment as Andrew Jackson's secretary of state. In 1832, Van Buren was Jackson's running mate; he was elected vice president. Riding on Jackson's coattails, Van Buren was easily elected president in 1836. Until George Bush's election in 1988, Van Buren was the only candidate who succeeded to the presidency from the vice presidency. The Panic of 1837, though, made him extremely unpopular. He was defeated by William Henry Harrison in 1840. In 1848, Van Buren ran for president as the Free Soil Party candidate. His presence on the ballot gave Zachary Taylor the victory.

Van Ness Street (NW)

Formerly Shepherd Street, the street was renamed in 1905 to honor the Van Ness family. Prior to her marriage to John Peter Van Ness (d. 1846), Marcia Burnes (1782–1832) was considered the "heiress of Washington City." Her farmer father, Davey Burnes, had substantial real estate holdings in what had become the capital city and, upon his death, Marcia inherited his property. John Peter Van Ness was a congressman from Massachusetts. Expelled from Congress for violating a Constitutional clause by accepting an appointment from Thomas Jefferson as major in the District Militia (no one may hold two positions simultaneously in the United States government), Van Ness moved his residence to Washington. He held positions as president of the Bank of the Metropolis and the Washington branch of the U.S. Bank, as well as retaining his militia assignment. In 1830, he became mayor of Washington. Marcia Burnes Van Ness put her time and effort into charity. In 1832, while aiding victims of a cholera outbreak, she fell ill and died of the disease. The day after her death, Congress adjourned to pay tribute to her. To this day, she is the only woman so honored by Congress.

Varnum Place (NE), Street (NE, NW)

James Mitchell Varnum (1748–1789) was known for his literary mode of speech, a trait that helped him become a successful lawyer. During the Revolutionary War, Varnum served at the Boston Seige before assuming commands that were not directly involved with the fighting. In 1777, Varnum took command of Forts Mercer and Mifflin.

Varnum thought the establishment of winter quarters at Valley Forge was ridiculous. An archeological dig at his Valley Forge headquarters, though, proved that Varnum made the best of the hardship. The dig uncovered hundreds of pounds of chicken bones. He finally saw action in the Monmouth Campaign and served under the Marquis de Lafayette at Newport. Varnum pressed for the creation of a Negro unit in Rhode Island; the force saw battle in late August 1778. In 1779, Varnum's forces mutinied. A few months after the mutiny,

Varnum resigned and was appointed Major General of the Rhode Island militia.

Varnum served in the Continental Congress (1780–82 and 1786–87) and in 1787 became a judge for the Northwest Territory, a region in which he had shown interest. Varnum was instrumental in the formation of a code of territorial laws.

Verbena Street (NW)
See Flora and Fauna.

Vermont Avenue (NW)
See State Streets.

Vernon Street (NW)
Vernon Street is named for Admiral Edward Vernon, commander of the British fleet in the Caribbean. George Washington's Virginia residence, Mt. Vernon, is also named for him. Washington's brother served under the commander.

Vine Street (NW)
See Flora and Fauna.

Virginia Avenue (NW, SE, SW)
In L'Enfant's original plan, Virginia Avenue was much shorter. Thomas Jefferson extended the street; he felt length was directly related to respect. (See State Streets.)

VIRGINIA AVENUE. Thomas Jefferson extended this diagonal street before the original city plan was completed. The Kennedy Center is located on the far left, along the Potomac River. The infamous Watergate Hotel rises on its right. The Francis Scott Key Bridge crosses the river in the distance. [CREDIT: ALA.]

Volta Place (NW)
The Georgetown Citizens' Association named Volta Place in 1905 as "a compliment to Alexander Graham Bell (1847–1922) in honor of winning the French government's Volta Prize." Bell used the prize money gained from his invention of the telephone to create the Volta Bureau (for the Increase and Diffusion of Knowledge relating to the Deaf). The bureau was first housed in the carriage house belonging to Bell's parents.

Wade Road (SE)
Benjamin F. Wade (1800–1878) served in the Senate as a Whig and a Republican. He strongly supported tough sanctions against the South after the Civil War. As president pro tempore of the Senate, Wade would have succeeded to the highest office if Andrew Johnson had been impeached. He was nicknamed "Bluff Ben" after he was challenged to a duel by a Southerner. As the challengee, he had the choice of weapons. He chose squirrel rifles at 20 paces; the challenger thought he was crazy and backed down.

Walnut Avenue (NW), Street (NE)
See Flora and Fauna.

Walter Street (SE)
About 20 years after it was built, the Capitol, designed by Dr. William Thornton, became too small for the number of state representatives serving in Congress. In 1851 Congress authorized a competition to determine how the building would be altered. Thomas U. Walter (1804–87) submitted the victorious design, which extended the wings and placed the first cast iron dome in the world on top. The modifications were completed in 1863. (See Thornton Place.)

Ward Circle (NW), Place (NW)
Major General Artemas Ward (1727–1800) commanded troops in three wars, but due to disagreements with

George Washington during the Revolutionary War, he eventually resigned his command. During the latter war, Ward, after rising from his sick bed, commanded forces at the Boston Seige until Washington arrived.

> "Perhaps he deserved more credit than he received. He kept the Army together in front of Boston until Washington came, and after that, however much he felt aggrieved, he did not add to his successor's difficulties by organizing the discontented. Charles Lee laughed at Ward's generalship and sneeringly termed him a church warden. Perhaps it was a compliment. Had Ward possessed the spirit of the man who assailed him, he might have ruined Washington."

Ward also served as a delegate to the Continental Congress (1777-1778) and as a judge.

Ward probably would not have received a statue in his honor had it not been for his great grandson, who left one million dollars to Harvard University, Ward's alma mater. Sound simple? The condition of the bequest was that Harvard erect and maintain a statue of Ward in the District of Columbia. The university kept its part of the bargain, and in 1927 offered fifty thousand dollars for the statue. Congress approved the project, but by 1938, the price of bronze had skyrocketed and General Ward had to be immortalized horseless. The statue stands in the center of Ward Circle at the intersection of Massachusetts and Nebraska Avenues.

Washington Avenue

One of the most popular trivia questions about the District concerns the forgotten state street, Washington. Unfortunately, the question, and its answer, became obsolete on November 16, 1989. A section of Canal Street (between South Capitol Street and Independence Avenue) was renamed for the state. The change was the result of a campaign launched by state resident C. S. Wetherell.

Washington Circle (NW)

Washington Circle connects Pennsylvania and New Hampshire Avenues, 23rd and K Streets and is named for George Washington (1732-99), the "father of our country," the first president of the United States and planner of the city of Washington. Washington died before the city became a reality in 1800. The statue in the center of the circle represents George Washington at the Battle of Princeton.

This circle was the site of the city's first race course.

Water Street (NW, SE, SW)
Water Street, which is aptly named, is divided into three parts. One section travels along the Washington Channel from M Place until it deadends near the Francis Case Memorial Bridge. Another segment, along the Anacostia River, links T Street and Potomac Avenue. The last part, which also travels along the Anacostia, runs between I-295 and I-395. The street, in existence prior to the American Revolution, is the oldest street in Washington.

Waterside Drive (NW)
Waterside Drive originally followed the Potomac River upstream to Rock Creek. Today, it is still accurately named, but only two disconnected blocks remain along Rock Creek.

Wayne Place (SE)
Wayne Place is named for Revolutionary War general Anthony Wayne (1745–1796). Nicknamed "Mad Anthony," he won recognition for capturing Stony Point, New York, during the war. After the revolution, Wayne moved west and fought in the Indian Wars.

Webster Street (NE)
Daniel Webster (1782–1852) won national recognition as a lawyer in the Dartmouth College case and *McCullouch v. Maryland*. He held many political positions, including two stints as Secretary of State, one under William Henry Harrison and the other under Millard Fillmore.

Wesley Circle (NW), Place (SW)
Wesley Circle is named for John Wesley (1703–1791), the founder of Methodism. Wesley preached over 40,000

sermons during his evangelical career. In 1784, he legally established Methodist societies. His brother, Charles (1707–1788), also a Methodist preacher, wrote 6,500 hymns, including "Hark! The Herald Angels Sing."

West Basin Drive (SW)
West Basin Drive travels along the western shore of the Tidal Basin.

West Beach Drive (NW), Terrace (NW)
See Beach Drive.

Western Avenue (NW)
The District of Columbia and Montgomery County, Maryland, are separated by Western Avenue. The street runs along Rock Creek Park from Westmoreland Circle, just short of the Dalecarlia Reservoir grounds, to Oregon Avenue.

West Executive Avenue (NW)
Like its sister street, East Executive Avenue, this street is, for the most part, inaccessible to the public. The street provides access to the Old Executive Office Building from the White House.

WEST BASIN DRIVE/OHIO DRIVE. West Potomac Park juts out along the Tidal Basin. The two streets that allow access to the park are West Basin Drive (left) and Ohio Drive. [CREDIT: ALA.]

Westminster Street (NW)
This street, located between 9th and 10th Streets, is one block long. The street is named for Westminster Abbey. The abbey, a national shrine of England, has been home to the coronations of all English monarchs since William I.

West Virginia Avenue (NE)
See State Streets.

Wheeler Road (SE)
Joseph "Fightin' Joe" Wheeler (1836–1906) unwittingly helped to close the chasm between the North and South after the Civil War. Wheeler's troops were the only Confederate troops to challenge Sherman's march to the sea, albeit unsuccessfully. After the war, he served 15 years in the House of Representatives. During the Spanish-American War, William McKinley appointed Wheeler major general of volunteers, making him the first Confederate to hold a post-Civil War command in the U.S. Army. He was later promoted to brigadier general in the regular army.

Whitehaven Parkway (NW), Street (NW)
Whitehaven was the name given by Amos Cloud to his property. He purchased the land from Benjamin King (see King Street). He later sold the house to Colonel Henry Carbery, who renamed it Cincinnati and then Carbery Glen. The original house now exists as the wing of a house, purchased by Nelson Rockefeller in the late 1940s, on 49th Street across from Battery Kemble Park.

Whitehurst Freeway
This thoroughfare connecting Georgetown and Foggy Bottom was completed in 1949. The road was created by an incredible union. The work was completed by a firm named Alexander and Repass. The remarkable part is that in this tense segregated era, Alexander was black and Repass was white. Their handiwork also includes the Tidal Basin.

Willard Street (NW)

He operated the first overnight delivery service, running packages between the two Hudson River ports, but Henry Willard is best recognized for the historic hotel that bears his name. Willard started at Fuller's City Hotel as its manager in 1847. He then leased the tarnishing hotel, located two blocks from the White House, from owner Benjamin Ogle Tayloe. Shortly after, he purchased the hotel and gave it his name.

His fine management skills and family help made the Willard Hotel one of the finest in the city, with visiting dignitaries clamoring to reserve its rooms. During the Civil War, the Willard accommodated both Northerners and Southerners. Guests were assigned floors and exits depending upon the side of the Mason-Dixon line from which they came. Nathaniel Hawthorne described the hotel: ". . . the Willard's Hotel could more justly be called the center of Washington and the nation than either the Capitol, or the White House, or the State Department." It is said that Ulysses S. Grant enjoyed sitting in the hotel lobby, a lit cigar in his mouth, watching the lobbyists work.

The Willard was the first hotel in America to provide its guests with desks. Willard also provided elaborate meals for his guests, believing that to be the way to the patrons' hearts and wallets.

In the late 1980s, the hotel was once again becoming run-down, but the historic landmark received a facelift and is again operating as one of Washington's luxury hotels.

WILLARD STREET. The Willard Hotel has been a stop for Washington politicians and socialites since Henry Willard began serving the city around 1948. [CREDIT: ALA.]

Williamsburg Lane (NW)
See City Streets.

Willow Street (NW, SE)
See Flora and Fauna.

Wilmington Place (SE)
See City Streets.

Windom Place (NW)
William Windom served as Secretary of the Treasury (1881, 1889-91) under James Garfield and Benjamin Harrison.

Wisconsin Avenue (NW)
Wisconsin Avenue, which rises from the Potomac and continues north through Maryland, predates the rectangular grid plan of Washington. It originated as an Indian trail and became a rolling road as commercial trade developed. During Colonial times, the name High Street graced part of the thoroughfare; farther north it was named Frederick Road. A century later it was called 32nd Street, and in 1906, Congress named it after the state.

In the 1700s, when Georgetown was a booming tobacco port, High Street served as the principal commercial route northwest to the Ohio Valley. During the French and Indian Wars, George Washington, supporting British General Edward Braddock's troops, twice (1753 and 1754) led soldiers along this route to Fort Duquesne (Pittsburgh).

Wisconsin Avenue is home to the National Cathedral (at Massachusetts Avenue) and the Soviet Embassy compound.

The Soviet compound has long been considered the highest point in the city, but it runs second to the bank of television towers at the corner of Brandywine Street.

Woodley Place (NW), Road (NW)
Woodley, a house located at 3000 Cathedral Avenue, served as the summer home for Presidents Martin Van Buren, John Tyler, James Buchanan and Grover Cleveland. The house, built by Philip Barton Key, was owned by Henry Stimson. In 1950, it became part of the Maret School, where it still stands.

Woodley Road runs perpendicular to the property on both sides; along with the surrounding neighborhood, it takes its name from the house.

Wyoming Avenue (NW)
In August 1905, Wyoming Senator Francis E. Warren registered his protest to the proposed changing of Wyoming Avenue to Winthrop Street. The senator stated that he had lived on the street named for the state he proudly represented for many years. The commissioners, responding to complaints from Warren and other residents, decided to allow the street name to stand. (See State Streets.)

Yorktown Road (NW)
Yorktown Road is named for the final campaign of the Revolutionary War. British General Cornwallis, having been defeated in the Carolinas, dug his troops in at Yorktown to await reinforcement. The reinforcement never showed and Cornwallis was forced to surrender (1781) to Generals Washington and Rochambeau and the Marquis de Lafayette. Cornwallis's surrender brought the rebellion to an end.

Yuma Street (NW, SE)
See City Streets.

Bridges

One noticeable feature of bridges leading to and from the District of Columbia is that none have overhead structures. Judging from current air traffic and flight patterns, this is a definite plus.

American Legion Bridge

The American Legion Bridge, which stretches 220 feet and rises 101 feet to the top, spans the Cabin John Creek. Formerly the Union Arch Bridge and the Cabin John Bridge, it was designed by Montgomery C. Meigs in 1857 to bring water from Great Falls to Washington. The bridge was considered one of the largest masonry arches in the world. Plaques are located on the bridge honoring Abraham Lincoln and Jefferson Davis. Davis's plaque is the only memorial to him in the District.

Anacostia Bridge

See 11th Street Bridge.

Arland D. Williams Jr. Bridge

Arland D. Williams Jr. Bridge, one of the 14th Street Bridges, is named to honor a hero and victim of a January 13, 1982, Air Florida plane crash that killed 74 passengers and four commuters on the bridge.

Arlington Memorial Bridge

The construction of Memorial Bridge was proposed by Daniel Webster in 1851 as a symbol of the firmly established Union. The start of the Civil War and the dissolution of the Union postponed all construction plans.

In 1901, the specific location and design were chosen by the Senate Park Commission. The final design, by architects McKim, Mead and White, was accepted by Congress in 1924. The bridge, according to the design, would serve as a monument, instead of a means to relieve traffic congestion. Construction took six years, requiring river channeling and reconstruction of the Mall. Officially opened in 1932, the bridge joined the Lincoln Memorial and Arlington National Cemetery, rejoining north and south. The bridge is equipped with a draw span designed by Joseph B. Strauss (of Golden Gate Bridge fame). The machinery, skillfully concealed under the center span, was operational until 1974.

Benning Road Bridge

The current bridge was constructed in 1934, although the site is one of the oldest in Washington.

Chain Bridge

In 1797 Falls Bridge was built; it was the first span across the Potomac. Prior to its construction, a ferry transported people from one side of the river to Analostan (Roosevelt Island). A toll of 10 pence was charged to help pay for the construction, and constant reconstruction, of the bridge. Since the bridge was made of wood, it was often washed away by high water and ice, so the cost of maintenance was exorbitant.

The wooden bridge was replaced by a chain bridge in 1810, thus the name. The current bridge is the eighth to span the 250-foot-deep Potomac River gorge.

Connecticut Avenue Bridge (Klingle Valley Bridge)

The Connecticut Avenue Bridge, which carries the street over Klingle Valley, is rarely recognized as a bridge by the pedestrians and motorists travelling through Cleveland Park . . . unless they happen to look over the railing. The Fine Arts Commission described the bridge in 1974 as "at once anonymous, proper and subservient to the city plan around it." The first bridge on this site was constructed in 1891 when Connecticut Avenue was extended into the "country." The bridge was primarily a streetcar bridge. The bridge that stands today, designed by Paul Cret, was built in 1932. Two urns are positioned at each end of the bridge.

Duke Ellington Bridge (Calvert Street Bridge)

Until 1907, the Calvert Street Bridge was the only upstream crossing of Rock Creek Park, joining the Adams Morgan neighborhood to Connecticut Avenue. The bridge was first constructed in 1891 to carry the street car line over the Rock Creek valley gorge. The current bridge was designed by Paul Cret in 1935. In an attempt to prevent the stranding of residents, the old bridge was relocated during construction of the new one. Horses dragged the old bridge, one section at a time, downstream 80 feet. Once the new bridge became operational, the old one was destroyed.

The bridge was renamed to honor Duke Ellington in 1975, shortly after his death.

Edward Kennedy Ellington (1899–1974), considered one of the greatest jazz musicians, was born in Washington, D.C. He acquired the nickname "Duke," a reference to the grace with which he carried himself, while a teenager. The many piano players in his neighborhood (his house was located at 1212 T Street, NW) and his mother, who also played, urged him to take lessons. He composed his first piece, "Soda Fountain Rag," while working at an ice cream parlor near Griffiths Stadium. He was 14.

In 1918, Ellington and his group made a name for themselves performing in Harlem clubs. By 1920, the Washingtonians were in demand in the District of Columbia. Ellington composed over 6,000 pieces, including "A-Train," "Stormy Weather," and "Sophisticated Lady." In 1943, Ellington gave the first non-classical performance at Carnegie Hall.

DUMBARTON BRIDGE. This bridge, constructed with a 12-degree curve, connects two unaligned sections of Q Street. The anatomically correct buffaloes, located on both sides of the bridge, are often the target of vandalism. [CREDIT: ALA.]

Dumbarton Bridge (Buffalo Bridge)

Completed in 1914, this bridge spans Rock Creek and Rock Creek Parkway and connects two sides of Q Street that are not aligned. The bridge, therefore, was built with a 12-degree curve. Designed by the father-son team of Glenn and Bedford Brown, this bridge features 28 Indian heads modeled after a life mask of Chief Kicking Bear and four buffaloes designed by A. Phimister Procter. The bridge is an imitation of the medieval Ponte Maggiore Bridge in Italy.

Eleventh Street Bridge (Officer Kevin J. Welsh Memorial Bridge)

A bridge on this site served as John Wilkes Booth's escape route from Washington after shooting Abraham Lincoln. The 11th Street bridge is a twin span with the Anacostia (12th Street) Bridge. Construction of these bridges across the Anacostia River made Anacostia the perfect location for the city's first blue collar neighborhood. The twins were built to join I-395 and I-295. The spans were built in 1965 and 1970.

Fourteenth Street Bridges

Three bridges, named Rochambeau Bridge, George Mason Memorial Bridge and Arland D. Williams Bridge, carry 14th Street over the Potomac River to I-395. The fourth in this cluster is a railroad bridge. The vehicle bridges were built in 1950, 1962 and 1971, respectively.

Francis Scott Key Bridge

Key Bridge was completed in 1923 on the site of an aqueduct bridge. An extra arch was added in 1939 to span the George Washington Parkway. The Aqueduct Bridge provided a commercial connection between the C&O Canal and Alexandria, Virginia. During the Civil War, planks were placed across the bridge to ease the transport of Union troops into Virginia. The government was extremely worried about the possibility of Confederate attacks on the federal city, therefore, the planks of the bridge were removed each evening and replaced the next morning. The supports of the old Aqueduct Bridge are still visible in the waters of the Potomac River.

The Three Sisters Bridge, started in 1972 slightly upriver from the Key Bridge, was never completed, although its foundations still remain on the Georgetown side of the river. It was destroyed by a flood, and some Washingtonians believe its failure was caused by an Indian curse.

The bridge's name honors Francis Scott Key, probably best known as the lyricist of "The Star Spangled Banner." The words were penned while Key was stuck at Fort McHenry during the British attack of September 12, 1814. Key had been at the fort to negotiate the release of an American prisoner held on a British ship. The negotiation was successful, but the British attacked before Key could depart. Unable to tell who held the fort until sunrise, Key was happy to see the American flag still flying over the fort. His lyrics, set to the music of "To Anacreon in Heaven," were adopted by Congress as the National Anthem in 1931. The anthem had already been adopted by the armed forces.

Key was a solid Georgetown citizen, maintaining a law practice as well as serving three terms as U.S. Attorney for the District of Columbia. Along with his eight children, he lived in a house that was located just west of the bridge. The house stood until 1949, when construction for the Whitehurst Freeway forced its demolition.

Frederick Douglass Memorial Bridge

The Douglass Bridge has the only drawbridge across the Anacostia River. The bridge was constructed in 1949 and named the South Capitol Street Bridge. The name was changed to honor abolitionist Frederick Douglass, whose residence is a national park in Anacostia, in 1965.

Harvard Street Bridges

These two bridges allow access to the National Zoological Park over Rock Creek. The older of the two, built in 1891 and reconstructed in 1901, enters the Zoo from Beach Drive. The second, constructed in 1965, enters from Harvard Street and Adams Mill Road.

George Mason Bridge

The bridge, one of the 14th Street Bridges, is named to honor George Mason (1725–92), author of the Virginia Declaration of Rights (1776). His Virginia declaration served as a form for the first section of the Declaration of Independence. Mason was active in drafting the Constitution, but he disagreed with some of the provisions and opposed its ratification. Although he opposed some aspects of the Constitution, Mason assisted with the creation of the Bill of Rights.

Charles C. Glover Bridge (Massachusetts Avenue Bridge)

This bridge over Rock Creek Park is named to honor Charles C. Glover (see Glover Road). The first bridge on this site was constructed in 1901, when Massachusetts Avenue was extended. The current bridge, built in 1941 to better accommodate those moving out of the city center, rises 75 feet above Rock Creek Park.

John Philip Sousa Memorial Bridge

The John Philip Sousa Memorial Bridge, completed in 1940, carries Pennsylvania Avenue across the Anacostia River.

Sousa (1854–1932), a Washington native, led the U.S. Marine Band for 12 years. Sousa composed over 100 marches, including "The Stars and Stripes Forever," "The Washington Post," and "Semper Fidelis."

K Street Bridge

The two-tiered K Street Bridge spans Rock Creek Park below the Whitehurst Freeway. The first bridge was constructed around 1793. Lots for buildings were created by the filling and draining of the area. Land was donated by the owners with the provision that the government construct a bridge. The government sold the property in an attempt to gain money for the bridge construction, but the revenues fell short and an appropriation had to be sought. The bridge was constructed with three arches. The center arch contained 13 stones, each one representing a state in the Union. The keystone represented Pennsylvania; it is thought that the state derived its nickname of the "Keystone State" due to this stone. The bridge was engraved: "May the Union last as long as this bridge."

The bridge was removed four years after its construction. It was not replaced until 1869. Another bridge was constructed in 1907. The current bridge was completed and opened in 1941.

Long Bridge

One of the 14th Street Bridges, Long Bridge was first constructed as a drawbridge in 1809. The bridge over the Potomac was often washed away by ice and flood waters. Today, the bridge carries railroad tracks from East Potomac Park to Arlington, Virginia.

M Street Bridge

The first bridge on this site was constructed in 1788, making it the oldest in the city. The bridge was built in line with M Street by the city of Georgetown two years prior to the creation of the capital. The bridge constructed on this site in 1800 was a drawbridge. The draw was needed to accommodate the sailing vessels that travelled Rock Creek and the Anacostia River. The first steel bridge was constructed in 1871; it was condemned in 1928. The current bridge was built in 1930 to span the newly constructed Rock Creek Parkway.

Local legend states that a stagecoach driver and four horses were lost when the bridge washed out during a violent storm. It is said that their ghosts can be seen crossing the bridge on stormy nights.

Military Road Bridge
This bridge carries Military Road across Rock Creek. The first bridge was constructed in 1862 by the Union Army to complete the transport road (see Military Road). The bridge was replaced in 1905. The present bridge was constructed in 1929.

P Street Bridge
This bridge carries P Street across Rock Creek at the site of an old ford. The ford existed until 1850. Generals Lafayette and Rochambeau led French units across this point in the creek heading north to Yorktown during the American Revolution.

Rochambeau Bridge
The bridge is named for Jean Baptiste Donatien de Vimeur, comte de Rochambeau (1725–1807), who landed at Newport, Rhode Island, in 1780 with French troops and, with George Washington, planned the victorious Yorktown Campaign. Rochambeau Bridge is one of the 14th Street Bridges.

Shoreham Bridge
This bridge crossing Rock Creek was first built in 1926. The steel trusses from the old Aqueduct Bridge were used in its construction. A wider bridge was constructed on the site in 1938.

16th Street Bridge
This bridge, built in 1910, spans Piney Branch Valley. The four tigers that guard its entrances were sculpted by A. Phimister Proctor.

Theodore Roosevelt Memorial Bridge

This bridge, built by the D.C. Highway Department to provide access to Virginia and Roosevelt Island, is a complex network of freeways (mainly Route 50 and I-66). The bridge, opened in 1964, provides the only vehicular access to Theodore Roosevelt Island. At the turn of the 20th century, a tunnel was proposed for this site.

Whitney M. Young Jr. Memorial Bridge

This bridge extends East Capitol Street across the Anacostia River and National Children's Island. The bridge was built in 1955.

Whitney M. Young, Jr. (1921–1971), served as head of the National Urban League for ten years, from 1961 to 1971. Trained as a social worker, Young was one of the leading civil rights activists of the 1960s.

William Howard Taft Bridge

The bridge was named for President William Howard Taft in 1931. Taft lived near the bridge and was often seen taking walks across it. After its construction, the bridge was named "The Million Dollar Bridge," due to its $846,000 price tag. This bridge, which carries Connecticut Avenue over Rock Creek Park, was constructed directly above the Woodley Lane Bridge (or High Bridge). Remnants of the earlier bridge can be seen below. Taft Bridge opened in 1907 after 10 sporadic years of construction. At that time, this was considered the largest concrete bridge in the world. Designed by George S. Morison, this bridge contains no metal reinforcement. The bridge is guarded by four lions designed by A. Phimister Proctor.

Woodrow Wilson Bridge

The Wilson Bridge is the only drawbridge in the nation's interstate system. It is also the longest bridge in Washington, D.C. (5,897 feet). The Wilson Bridge, which carries the Beltway (I-495) over the Potomac, was built in 1961.

Bibliography

Alotta, Robert I. *Civil War Justice: Union Army Executions Under Lincoln.* Shippensburg, PA: White Mane Publishing Co., Inc., 1989.

Alotta, Robert I. *Mermaids, Monasteries, Cherokees and Custer.* Chicago: Bonus Books, Inc., 1990.

Applewhite, E.J. *Washington Itself.* New York: Alfred A. Knopf, 1981.

Arnebeck, Bob. *Through a Fiery Trial Building Washington 1790–1800.* Lonham, MD: Madison Books, 1991.

Billings, Mrs. Elden E. "Alexander Robey Shepherd and His Unpublished Diaries and Correspondence." *Records of the Columbia Historical Society of Washington, D.C.,* 1960–62.

Bisbort, Alan. "The Draw of Bridges." *The Washington Post, Weekend Section,* 10 April 1992.

Boatner, Mark M., III. *Encyclopedia of the American Revolution.* New York: David McKay Co., 1966.

Brinkley, David. *Washington Goes to War.* New York: Ballantine Books, 1988.

Brown, George Rothwell. *Washington: A Not Too Serious History.* Baltimore: The Norman Publishing Co., 1930.

Bryan, Wilhelmus B. "Some Myths in the History of Washington." *Records of the Columbia Historical Society of Washington, D.C.,* 1930.

Bryant, Wilhelmus Bogart. *A History of the National Capital.* New York: The MacMillan Company, 1914.

Castle, Guy. "Blue Plains and Bellevue: Two Early Plantations of the Washington Area." *Records of the Columbia Historical Society of Washington, D.C.,* 1953–55.

Castle, Guy. "The Washington Area Between 1608 and 1708, with a Biographical Note on Prince George of Denmark." *Records of the Columbia Historical Society of Washington, D.C.,* 1963–65.

Bibliography

Clark, Allen C. "The Old Mills." *Records of the Columbia Historical Society of Washington, D.C.*, 1930.

Clark, Allen C. *Origin of the Federal City*. DC: Columbian Historical Society, 1935.

Commemorative Program. "The Battle of Fort Stevens." July 15-16, 1989.

Cooling III, Benjamin Franklin and Walton H. Owen II. *Mr. Lincoln's Forts: A Guide to the Civil War Defenses of Washington*. Shippensburg, PA: White Mane Publishing Co., Inc., 1988.

Emery, Fred A. "Mount Pleasant and Meridian Hill." *Records of the Columbia Historical Society of Washington, D.C.*, 1932.

Emery, Fred A. "Washington's Historic Bridges." *Records of the Columbia Historical Society of Washington, D.C.*, 1938.

Engineering Department files. District of Columbia Archives, 1898-1908.

Eskew, Garnett Laidlaw. *Willard's of Washington*. New York: Coward-McCann, Inc., 1950.

Evelyn, Douglas E. and Paul Dickson. *On This Spot: Pinpointing the Past in Washington, D.C.* Washington, D.C.: Farragut Publishing Company, 1992.

Ewing, Charles. *Yesterday's Washington, D.C.* Miami, FL: E.A. Seemann Publishing, Inc., 1976.

Faust, Patricia L., ed. *Historical Times Illustrated Encyclopedia of the Civil War*. New York: HarperPerennial, 1986.

Fohr, Stephen C. "On Freeway, Hardly a Sign of Renaming." *Washington Post*, 18 January 1992.

Fogle, Jean. *Two Hundred Years Stories of the Nation's Capital*. Arlington, VA: Vandamere Press, 1991.

Fox, Larry. "Naming Names." *Washington Post Weekend*, 3 January 1992.

Fronook, Thomas, ed. *The City of Washington*. New York: Alfred A. Knopf, 1977.

Green, Constance McLaughlin. *Washington Village and Capital. 1800-1878*. Princeton, NJ: Princeton University Press, 1962.

Green, Constance McLaughlin. *Washington: A History of the Capital 1800-1950*. Princeton, NJ: Princeton University Press, 1962.

Gutheim, Frederick. *The Federal City: Plans and Realities*. Washington, DC: Smithsonian Institution, 1976.

Hilzenrath, David S. "Historic House Sold at Foreclosure Auction." *Washington Post*, 5 April 1991.

Jackson, Cordella. "People and Places in Old Georgetown." *Records of the Columbia Historical Society of Washington, D.C.*, 1932.

Bibliography

Kenny, Hamill. *The Placenames of Maryland, Their Origin and Meaning.* Baltimore: Maryland Historical Society, 1984.

Kite, Elizabeth S. *L'Enfant and Washington.* Baltimore: The Johns Hopkins Press, 1929.

Kohler, Sue A. and Jeffrey R. Carson. *Sixteenth Street Architecture.* Washington, D.C.: U.S. Government Printing Office.

Larner, John B. "Some Reminiscences of Mrs. John M. Binckley of Early Days in Washington." *Records of the Columbia Historical Society of Washington, D.C.,* 1928.

Lewis, David L. *District of Columbia, A History.* New York: W.W. Norton & Co., Inc., 1975.

Martin, Edward Winslow. *Behind the Scenes in Washington, DC:* Continental Publishing Company and National Publishing Company, 1873.

Morris, Richard S., ed. *Encyclopedia of American History.* New York: Harper and Row Publishers, 1976.

Newell, Frederick Haynes, ed. *Planning and Building the City of Washington, DC:* Ransdell Inc., 1932.

Nicolay, Helen. *Our Capital on the Potomac.* New York: The Century Co., 1924.

Proctor, John Clagett. *Proctor's Washington and Environs.* Washington: written for *The Washington Sunday Star,* 1928–49.

Proctor, John Clagett. *Washington Past and Present.* New York: Lewis Historical Publishing Company, Inc., 1930.

Spratt, Zack. "Rock Creek's Bridges." *Records of the Columbia Historical Society of Washington, D.C.,* 1953.

"Street Name to Mark Freedom." *Washington Post.* 4 July 1991.

Urdang, Laurence. *Names and Nicknames of Places and Things.* New York: Meridian, 1987.

Van Doren, Charles, ed. *Webster's American Biographies.* Springfield, Massachusetts: G. & C. Merriam Company, 1975.

Weller, M.I. "Commodore Joshua Barney: The Hero of the Battle of Bladensburg." *Records of the Columbia Historical Society of Washington, D.C.,* 1911.

Worth, Fred L. *Strange and Fascinating Facts About Washington, DC.* New York: Bell Books, 1988.

Index

A

1st Street, 1
2nd Street, 36, 102
3rd Street, 17, 102
4½ Street, 69
4th Street, 90, 102
5th Street, 16, 102
6th Street, 16, 102
7th Street, 72, 94, 102
8th Street, 102
9th Street, 16, 102
10th Street, 101
11th Street, 101
11th Street Bridge, 154
12th Street, 101, 106
13th Street, 16, 50, 64, 101
14th Street, 101
14th Street Bridges, 152, 154
15th Street, 46, 52, 120
16th Street, 101, 129
16th Street Bridge, 158
17th Street, 29, 30, 101
18th Street, 17, 101
19½ Place, 101
19th Street, 103
20th Street, 101
23rd Street, 102
24th Street, 64
25th Street, 38
31st Street, 102
32nd Street, 102
33rd Place, 102
33rd Street, 102
34th Place, 102
34th Street, 52
37th Street, 102
38th Street, 102
63rd Street, 129
Adams, John, 3, 69, 83, 109
Adams, John Quincy, 3, 19, 26, 67
Adams Mill Road, 3, 102
Adams Place, 3
Adams Street, 3, 102
Akron, Ohio, 25
Akron Place, 3, 25
Alabama Avenue, 3, 49, 131
Alaska Avenue, 3, 131
Albany Street, 3, 102
Alden, James, 4
Alden Place, 4
Allison Street, 102
Alton Place, 102
American Fur Company, 5
American Legion Bridge, 151
American University, The, 83
Ames, Adelbert, 3
Ames Place, Street, 4
Anacostia, 4
Anacostia Avenue, Freeway, Road, 4
Anacostia Bridge, 151
Anacostia River, 4, 20, 29, 156
Andre, John, 5
Andrews, Frank M., 4
Andrews Air Force Base, 4
Andrews Circle, 4
Apple Road, 5, 47
Aqueduct Bridge, 158
Arcadia, 5
Arcadia Place, 5
Arden, 5
Arden Drive, 5
Argonne Forest, 5
Argonne Place, 5
Argyle Street, 101
Argyle Terrace, 5
Arizona Avenue, 5, 131
Arkansas Avenue, 6, 131
Arland D. Williams Bridge, 151, 154
Arlington Memorial Bridge, 151
Arlington National Cemetery, 78, 90
Armes Place, 102
Arnold, Benedict, 6
Arnold Avenue, Drive, 6
Arthur, Chester A., 56, 59
Asbury, Francis, 6
Asbury Park, N.J., 6

Index

Asbury Place, 6
Ashby, Turner, 6
Ashby Street, 6, 102
Aspen Street, 6, 47, 102
Astor, John Jacob, 7
Astor, John Jacob IV, 7
Astor, William Waldorf, 7
Astor Place, 6
Atlantic Ocean, 7
Atlantic Street, 7
Audubon, John James, 7
Audubon Terrace, 7
Austin Place, 108
Austin Road, 7
Austin Street, 104
Avon Lane, Place, 7
Avon River, 7
Azalea Road, 7, 48

B

Bacon, Henry, 8
Bacon Drive, 8
Baker, Edward Dickinson, 49
Baltimore Street, 16, 102
Bancroft, Edward, 33
Bancroft, George, 8
Bancroft Place, Street, 8
Bangor, Maine, 25
Bangor Street, 8, 25
Bank Street, 8
Banneker, Benjamin, 8, 41
Banneker Drive, 8
Barber, Mrs. M.C., 99
Barlow, Joel, 70
Barnes, Joseph K., 9
Barnes Street, 9
Barney, Joshua, 28
Barron, James, 33
Barry, John, 9
Barry Place, Road, 9, 102
Basin Drive, 10
Bataan, 10, 30
Bataan Street, 10
Bates, Edward, 10
Bates Road, Street, 10
Battery Kemble, 10
Battery Place, 10
Beach, Lansing H., 10
Beach Drive, 10
Beall, Ninian, 36
Beau Plaine, 15
Beech Street, 10, 47
Beecher, Henry Ward, 11
Beecher Street, 11
Beechwood Road, 11, 48
Bellair, Maryland, 11
Bellair Place, 11
Bellevue, 11
Bellview, 11
Bellview Drive, 11
Belmont Road, Street, 11, 102
Belt, Joseph, 11, 24
Belt Lane, Road, 11

Bending Lane, 12
Benning, Henry L., 12
Benning Road, 12
Benning Road Bridge, 152
Berkley, California, 25
Berkley Terrace, 12, 25
Berry Road, 12, 47
Bew Playnes, 15
Biltmore Street, 102
Bingham, Theodore A., 13
Bingham Drive, 13
Binney Street, 74
Birch Drive, Street, 13, 47
Birney, James Gillespie, 13
Birney Place, 13
Bismark Street, 104
Bladen, Thomas, 13
Bladensburg, Maryland, 13, 25, 28, 33, 48
Bladensburg Road, 13, 25, 50
Bladensburg Turnpike, 50
Blagden, Thomas, 5, 14
Blagden Avenue, Terrace, 14
Blaine, James G., 14, 26, 38
Blaine Mansion, 14
Blaine Street, 14
Blair, Francis Preston, 14
Blair Road, Portal, 14
Blanchard, Albert G., 15
Blanchard Drive, 15
Blue Plains Drive, 15
Blue Plains Sewage Disposal Plant, 11, 15
Bolling Air Force Base, 5, 11, 33, 41, 57, 74, 87
Booth, John Wilkes, 58, 79
Boundary Street, 108
Bowen, John Stevens, 15
Bowen Road, 15
Brads Avenue, 106
Brandywine Creek, Pennsylvania, 15
Brandywine Place, Street, 102, 104
Breeds Terrace, 106
Brent, Robert, 16
Brentwood, 16
Brentwood Parkway, Road, 16
Bridges, 151-9
Brightwood Avenue, 97
Brightwood Park Citizen's Association, 82
Broad Branch Road, 16
Bryant Street, 16, 102
Bryce, James, 88
Buchanan, James, 16, 35
Buchanan Street, 16, 102
Buchignani, Antonio, 40
Buckeye Drive, 17, 47

C

Cabin John Bridge, 151
Cabin John Creek, 151
California Avenue, 108

Index

California Street, 18, 131
Calvert, Cecilius, 19
Calvert, Charles, 19
Calvert, George, 19
Calvert, Leonard, 19
Calvert Street, 19, 102
Calvert Street Bridge, 153
Cambridge, Maryland, 25
Cambridge Place, 25, 34, 102
Camden, New Jersey, 25
Camden Street, 19, 25
Cammack Avenue, 107
Campbell, W. N., 97
Canal Road, 19, 22
Canal Street, 20
Capitol, 1, 90, 111
Capitol Heights, Maryland, 39
Carroll, Charles, 22
Carroll, Charles III of Carrollton, 20
Carroll, Charles of Bellevue, 109
Carroll, Charles of Carrollton, 20
Carroll, Daniel, 36, 78, 96
Carroll, Daniel of Carrollton, 20
Carroll, Daniel of Duddington, 20, 109
Carroll Avenue, 103
Carroll Row, 20
Carroll Street, 20, 103
Carrollburg Place, 20
Carrollsburg, 20
Carter, Jimmy, 109
Cathedral Avenue, 20, 103, 106
Catholic University of America, 61
Caton, Richard, 22
Caton Place, 22
Catonsville, Maryland, 22
Cedar Drive, Street, 22, 39, 47, 103
Central Avenue, 101
Chain Bridge, 22, 152
Chain Bridge Road, 22, 83
Champlain Avenue, 103
Champlain Street, 22, 103
Channing, William Ellery, 23
Channing Street, 23, 103
Chanute, Octave, 23
Chanute Place, 23
Chapin Street, 23
Charles C. Glover Bridge, 156
Charleston, South Carolina, 25
Charleston Terrace, 23, 25
Charlestown, Massachusetts, 13
Chase, Salmon P., 23
Chase Circle, 23
Cherry Road, Street, 24, 47
Chesapeake and Ohio Canal, 19
Chesapeake Bay, 24
Chesapeake Street, 24, 103, 104
Chestnut Street, 24, 39, 47, 103
Chevy Chase, 12, 24
Chevy Chase Circle, Parkway, 24
Chevy Chase Land Company, 24
Chicago, Illinois, 25

Chicago Street, 24
Church, B.S., 24
Church Street, 24
Cincinnati Street, 23, 102, 103
Clagett, Bishop, 22
Clagett, John, 26
Clagett Place, Street, 26, 101
Clagett, Thomas, 26
Clagett's Delight, 26
Claiborne, William, 18
Clark Place, 103, 105
Clay, Henry, 26, 67
Clay Place, Street, Terrace, 26
Claygate, England, 26
Clermont, 26, 53
Clermont Drive, 26
Cleveland, Grover, 14, 26, 59, 62, 75, 87, 95
Cleveland Avenue, 26
Cleveland Park, 73
Cleveland Street, 103
Cliffbourne Place, 27, 103
Cliffburn Place, 27
Clinton, George, 27, 68
Clinton Avenue, 103
Clinton Street, 27, 103
Cockburn, George, 54
Colfax Street, 104
College Place, 103
College Street, 27, 103
Colorado Avenue, 27, 131
Columbia Avenue, 101
Columbia Heights, 28, 62
Columbia Heights Citizen's Association, 109
Columbia Road, 28, 103, 138
Columbia Street, 103
Columbia University, 28
Columbus Street, 101
Concord Street, 66, 104
Conduit Road, 84
Congress Court, Place, Street, 28
Conifer Road, 29, 48
Conkling, Roscoe, 43
Connecticut Avenue, 10, 29, 103, 131
Connecticut Avenue Bridge, 152
Constitution Avenue, 17, 20, 65
Corcoran, William Wilson, 29, 61
Corcoran and Riggs Bank, 30
Corcoran Gallery of Art, 29
Corcoran Street, 29
Corregidor, 31
Corregidor Street, 10, 31
Cottage Place, 103
Crabtree Road, 31, 48
Crescent Place, 31, 103
Crescent Street, 103
Crittenden, John J., 31
Crittenden Street, 31, 103
Cromwell, Oliver, 31
Cromwell Terrace, 31
Cumberland, Maryland, 25
Cumberland Street, 25, 32

Index

Cushing, Caleb, 32
Cushing Place, 32, 103
Cutts, Richard, 75
Cutts-Madison House, 75
Czolgosz, Leon, 89

D

Dahlia Street, 32, 47, 103
Dana Place, 103
Danbury Street, 32
Dartmouth College, 75
Dartmouth Street, 75, 105, 138
Davenport, Thomas, 32
Davenport Street, 32, 103
Davis, Jefferson, 43, 90
Deane, Silas, 32, 74
Deane Avenue, 32
Dearborn Place, 104
Decatur, Stephen, 33, 75
Decatur Place, Street, 33, 103
Defense Boulevard, 33
Delafield, Richard A., 34
Delafield Place, Street, 34
Delaware Avenue, 34, 131
Dennison Place, 103
Dent, Frederick T., 34
Dent, Thomas, 57
Dent Place, 34
Denver, Colorado, 25
Denver Street, 25, 34
Detroit Street, 35, 103
Dewey, George, 22
Dexter, Samuel, 34
Dexter Street, 34
Dix, John Adams, 34
Dix Street, 34
Dogwood Street, 35, 47
Douglas, Stephen A., 35, 79
Douglas Street, 35, 103
Douglass, Frederick, 35, 36, 53, 156
Dover Street, 105
Downing, Andrew Jackson, 36
Downing Place, 36
Duddington, 36
Duddington Place, 36
Duke Ellington Bridge, 153
Dumbarton Bridge, 154
Dumbarton Oaks, 36
Dumbarton Rock Court, Street, 36
Dumbartonshire, Scotland, 36
Dunbar, Paul Lawrence, 37
Dunbar Road, 37
Duncan Street, 101
Dupont Circle, 14, 37, 50
duPont, Samuel Francis, 37
Dupont Street, 37

E

Eads, James Buchanan, 12, 38
Eads Street, 38
Eagle Nest Road, 39

Early, Jubal A., 50
East Avenue, 103
East Capitol Street, 39
East Executive Avenue, 39
East Potomac Park, 17
Eaton, John, 39
Eaton Place, 39
Eckington, 40
Eckington and Soldier's Home Railway, 40
Eckington Place, 40
Edgevale Terrace, 40
Edmunds, George F., 41
Edmunds Street, 41
Eglin Air Force Base, 41
Eglin Way, 41
Eisenhower, Dwight D., 5
Elder Street, 41, 47
Ellicott, Andrew, 9, 41
Ellicott, George, 9
Ellicott family, 8
Ellicott Street, Terrace, 41, 103
Ellington, Edward Kennedy, 153
Ellipse, 42
Ellipse Road, 42
Elm Street, 42, 47
Elmira, New York, 25
Elmira Street, 25, 42
Embassy Park Drive, 43
Embassy Row, 88
Emerson, Ralph Waldo, 43
Emerson Street, 43, 103
Emory Street, 104
Emporia Street, 43, 104
Erie Street, 43, 103
Eslin Avenue, 101
Euclid, 43
Euclid Place, 103
Euclid Street, 43, 103
Evarts, William M., 43
Evarts Place, 103
Evarts Street, 43, 103
Everett, Edward H., 43
Everett Street, 43
Executive Avenue, 129
"Eye" Street, 1

F

Fairmont Street, 44, 103
Faraday, Michael, 45
Faraday Place, 45
Farragut, David Glasgow, 3, 45
Farragut Place, Street, Square, 45
Farragut Street, 104, 105
Fenton, Reuben E., 45
Fenton Court, Place, 45
Fenwick, John, 45
Fenwick Street, 45
Fern Place, Street, 46, 47
Fessenden, William P., 46
Fessenden Place, 104
Fessenden Street, 46, 49, 104
Fillmore, Millard, 43

168

Index

Fitch, John, 46
Fitch Place, Street, 46
Flint, Michigan, 25
Flint Place, 25, 46
Flint Street, 82, 104
Floral Street, Terrace, 47, 48
Florida Avenue, 16, 48, 131
Folsom Street, 102
Foote, Andrew H., 12
Foote, Henry S., 48
Foote Street, 48
Forest Lane, 47, 49
Forrest, Uriah, 118
Forsyth Avenue, 106
Fort Baker, 49
Fort Baker Drive, 49
Fort Benning, Georgia, 12
Fort Davis, 50
Fort Davis Place, 50
Fort Dix, New Jersey, 34
Fort Drive, Place, 49, 104
Fort Dupont, 50
Fort Dupont Drive, Terrace, 50
Fort Dupont Park, 57
Fort Lesley J. McNair, 20
Fort Lincoln, 50
Fort Lincoln Drive, 50
Fort Mahan, 88
Fort Massachusetts, 50
Fort Pennsylvania, 90
Fort Reno, 49, 120
Fort Stanton, 49
Fort Stevens, 50, 84, 92
Fort Stevens Drive, 50
Fort Street, 107
Fort Sumner, 92
Fort Totten, 51
Fort Totten Drive, 51
Foxall, Henry, 51
Foxhall Road, 51, 104
Francis Scott Key Bridge, 155
Frankfort, Kentucky, 25
Frankfort Street, 25, 52, 103, 104
Franklin, Benjamin, 33, 52, 55
Franklin Street, 44, 104
Frederick, Maryland, 25
Frederick Douglass Court, 53
Frederick Douglass Memorial Bridge, 156
Frederick Place, 25
French, Daniel Chester, 38, 53
French Drive, 53
Friendship Heights, 12
Fuller, Margaret, 43
Fuller Street, 104
Fulton, Robert, 26, 46, 53, 80
Fulton Place, Street, Terrace, 53, 104

G

Gainesville, Virginia, 25
Gainesville Street, 25, 53
Galen, 53
Galen Street, 53
Galena, 54
Galena Place, 54, 104
Gales, Joseph, 54
Gales Place, Street, 54
Gallatin, Albert, 54
Gallatin Place, Street, 54, 104
Gallaudet, Edward, 55, 72
Gallaudet, Thomas H., 54
Gallaudet Street, 54
Gallaudet University, 55, 72
Galloway, Joseph, 55
Galloway Street, 55
Galveston, Texas, 25
Galveston Place, Street, 25, 56, 104
Garfield, James A., 9, 14, 41, 56
Garfield Street, Terrace, 56, 104
Garrison, William Lloyd, 56, 71
Garrison Street, 56
Gass, Stuart J., 109
Gates, Horatio, 56
Gates Road, 56
Gennessee Street, 106
George Mason Memorial Bridge, 154, 156
George Washington Memorial Parkway, 19
Georgetown, 1, 19, 83
Georgia Avenue, 57, 97, 131
Geranium Court, Street, 47, 57
Gerry Street, 104
Giesboro Road, 57
Giesborough, 57
Girard Street, 104
Gisbrough, 57
Glover, Charles Carroll, 57, 156
Glover Drive, Road, 57
Glover-Archbold Park, 57
Good Hope Road, 49, 58
Grant, Ulysses S., 15, 34, 58, 77, 81, 90
Grant Avenue, 102
Grant Circle, Road, 58
Grant Street, 58, 102, 104, 105
Greene, Nathanael, 58
Greene Place, 58
Greenleaf Street, 47, 58
Greenvale Street, 47, 58
Greenwich, Connecticut, 25
Greenwich Parkway, 25, 58
Gresham, Walter Q., 59
Gresham Place, 59, 104
Guiteau, Charles J., 56

H

Hadfield, George, 59
Hadfield Lane, 59
Half Street, 59
Hall, Asaph, 59, 99
Hall Place, 59

169

Index

Halley, Edmund, 60
Halley Street, Terrace, 60
Hamilton, Alexander, 60
Hamilton Circle, 60
Hamilton Street, 60, 104
Hamlin, Hannibal, 46, 60
Hamlin Place, Street, 60, 104, 105
Hancock Avenue, 103, 105
Hanna, Marcus Alonzo (Mark), 61, 89
Hanna Place, 61
Harewood, 61
Harewood Avenue, 102
Harewood Road, 61
Harlan, James, 61
Harlan Place, 61
Harris, Patricia Roberts, 109
Harrison, Benjamin, 14, 26, 61
Harrison, William Henry, 14, 30, 61
Harrisonburg, Virginia, 6
Hartford, Connecticut, 25
Hartford Street, 25, 62, 104
Harvard Street, Court, 62, 104, 138
Harvard Street Bridges, 156
Harvard University, 62
Hawaii Avenue, 62, 131
Hawthorne, Nathaniel, 62
Hawthorne Drive, Lane, Place, Street, 62, 104
Hayes, Rutherford B., 43, 62, 99
Hayes Street, 62
Hemlock Street, 47, 63
Henderson, John B., 91
Henry Bacon Drive, 63
Heurich, Christian, 38
Highland Avenue, 104
Highland Road, 104
Hoban, James, 63
Hoban Road, 63
Hobart, Garret A., 63
Hobart Place, Street, 63, 104
Holbrook Terrace, 107
Holly Spring Road, 47, 64
Holly Street, 47, 64
Holmead, Anthony, 64
Holmead, James Jr., 64
Holmead Avenue, 104
Holmead Place, 104
Holmes, Oliver Wendell, 50
Hopkins, Johns, 64
Hopkins Street, 64
Hospital Road, 64
Howard, Oliver O., 64
Howard Avenue, 104, 106
Howard Place, Road, Street, 64, 104
Howard University, 27
Howe, Sir William, 15, 17, 55
Howells, William Dean, 37
Hubbard Place, 102
Huntington Place, 104
Huron Street, 105

Hurst, John Fletcher, 65
Hurst Terrace, 65

I

I Street, 1
Idaho Avenue, 60, 65
Illinois Avenue, 65, 131
Independence Avenue, 20, 53, 65
Indiana Avenue, 66, 83
Indianapolis Street, 105
Ingle, Harry, 66
Ingleside, 66
Ingleside Terrace, 66
Ingraham, Joseph Holt, 66
Ingraham Street, 66, 104, 106
Iowa Avenue, 66, 131
Iris Street, 47, 66
Irving, Washington, 66
Irving Place, 34, 108
Irving Street, 66, 103, 105, 108
Itaska Street, 106
Ivy Street, 47, 67

J

J Street, 1
Jackson, Thomas Jonathan (Stonewall), 89
Jackson, Andrew, 14, 39, 67, 71, 92
Jackson Street, Place, 67
Jackson Street, 105
Jamaica Street, 67
Jay, John, 1, 67
Jay Street, 1, 67
Jefferson, Thomas, 16, 51, 63, 68, 86
Jefferson Drive, Place, Street, 68, 105
Jenifer, Daniel of St. Thomas, 69
Jenifer Street, 69
Jewett Street, 105
John Marr Circle, 69
John Marshall Place, 69
John Philip Sousa Memorial Bridge, 156
Johns Hopkins University, 64
Johnson, Andrew, 18, 41, 43, 46, 60, 69
Johnson Avenue, 69
Joliet, Illinois, 25
Joliet Street, 25, 70, 105
Jonquil Street, 47, 70
Judiciary Square, 69, 83
Juniata Street, 106
Juniper Street, 47, 70, 102
Justice Court, 70

K

K Street, 1
K Street Bridge, 157
Kalm, Peter, 70

Kalmia Road, 47, 70
Kalorama, 64
Kalorama Avenue, 105
Kalorama Road, 70, 105
Kanawha Street, 25, 71
Kanawha, West Virginia, 25, 71
Kansas Avenue, 71, 131
Kansas Street, 106
Kearny Street, 105
Keller, Helen, 22
Kendall, Amos, 55, 71, 94
Kendall Street, 71
Kenesaw Avenue, 105
Kenesaw Street, 105
Kenilworth Aquatic Gardens, 87
Kennedy, John F., 112
Kennedy Street, 105
Kensington, Maryland, 25
Kensington Place, 25, 72
Kentucky Avenue, 72, 131
Kenyon College, 72
Kenyon Street, 72, 105, 138
Keokuk Street, 105
Key, Francis Scott, 155
Kilbourne Place, 105
King, Martin Luther, Jr., 88
King, Rufus, 72
King Place, 72, 105
King Street, 106
Kirby, Edmund, 73
Kirby Street, 73
Kitchen Cabinet, 72
Klingle Road, 73
Klingle Street, 105
Klingle Valley Bridge, 152
Knox, Henry, 73
Knox Circle, Place, Street, Terrace, 73
Knox's "Noble Train of Artillery," 73

L

L'Enfant, Pierre Charles, 30, 36, 41, 77, 81, 96, 111
L'Enfant Promenade, Plaza, 77
Laboratory Road, 74
Lackland Way, 74
Lafayette, Marquis de, 9, 74
Lafayette Avenue, 74
Lafayette Square, 33, 75
Lake Erie, 43
Lamar Place, 107
Lamont, Daniel S., 75
Lamont Street, 75, 105
Langley, Samuel, 76
Langley Court, 76
Langley Way, 76
Lanier, Sidney, 76
Lanier Avenue, 105
Lanier Place, 76, 105
Lansing Street, 106
Larch Street, 102
Latrobe, Benjamin, 16, 20, 29

Laurel, Maryland, 25
Laurel Avenue, 105
Laurel Street, 25, 47, 76, 105
Lawrence, James, 76
Lawrence Avenue, Street, 76, 105, 107
Lawrence Place, 103
LeDroit Avenue, 102
Lee, Arthur, 33
Lee, Robert E., 77
Lee Street, 77
Lettered streets, 1
Lexington, Virginia, 25
Lexington Place, 25, 79
Library Court, 79
Library of Congress, 79
Lincoln, Abraham, 9, 10, 14, 24, 35, 37, 43, 49, 50, 58, 60, 61, 70, 79
Lincoln Drive, Road, 79
Lincoln Memorial, 8, 53, 98
Lincoln Street, 104
Lincoln-Douglas Debates, 35
Linden Place, 47
Linden Street, 102
Lingan, Nicholas, 109
Linnaeus, Carolus, 80
Linnean Avenue, 80
Linnean Hill, 80
Linthicum Place, 107
Little Falls Street, 80
Livingston, Edward, 81
Livingston, Philip, 81
Livingston, Robert R., 80
Livingston, Robert R., Jr., 80
Livingston, William, 81
Livingston Road, Street, Terrace, 80
Locust Road, 47, 81
Logan, John A., 81
Logan Circle, 81
Long Bridge, 24, 157
Longfellow, William Wadsworth, 82
Longfellow Street, 82, 106
Loring, William W., 82
Loring Way, 82
Loughboro Road, 65, 83, 106
Loughborough, Hamilton, 83
Loughborough, Nathan, 83
Louisiana Avenue, 66, 83, 131
Lover's Lane, 83, 102
Lowell, Charles R., 84
Lowell Lane, Street, 84, 106
Ludlow Avenue, 105
Luray, Virginia, 25
Luray Place, 25, 84
Luzon Avenue, 84
Lydecker Avenue, 106

M

M Street, 106
M Street Bridge, 157

Index

MacArthur, Douglas, 84
MacArthur Boulevard, 83, 84
MacArthur Terrace, 86
Macomb, Alexander, 86
Macomb Street, 43, 86
Madison, Dolly, 75
Madison, James, 27, 71, 86, 93
Madison Court, Drive, Place, Street, 86, 106, 107
Magazine Road, 87
Magnolia Street, 39, 47, 87
Maine Avenue, 87, 131
Mall, 13, 65, 76, 92
Manning, Daniel, 87
Manning Place, 87
Maple Avenue, 106
Maple Street, 47, 87, 106
Mapleview Place, 47, 87
Marine Place, 87
Marion, Francis, 87
Marion Street, 87
Marne Place, 87
Marne River, 87
Marr, John Q., 69
Marshall, John, 69
Marshall Street, 105
Martin Luther King Jr. Avenue, 88
Maryland Avenue, 88, 131
Mason, George, 110
Massachusetts Avenue, 43, 60, 65, 83, 88, 131
McClellan, George, 84
McClellan Avenue, 105
McClellan Street, 105
McGuire, Hunter H., 89
McGuire Avenue, 89
McKinley, William, 61, 63, 89
McKinley Place, Street, 89
McPherson, James B., 81
Meade, George Gordon, 90
Meade Street, 90
Meadow Road, 64
Meigs, Montgomery C., 30, 90
Meigs Place, 90, 106
Mellon, Andrew W., 91
Mellon, Richard, 91
Mellon Street, 91
Meridian Avenue, 106
Meridian Hill Park, 31
Meridian Place, 106
Messmore Place, 106
Messmore Street, 106
Michigan Avenue, 92, 131
Michler Street, 106
Military Road, 92, 107
Military Road Bridge, 158
Mill Road, 92
Mills, Robert, 92
Mills Avenue, 92
Milwaukee Street, 106
Minnesota Avenue, 92, 131
Mississippi Avenue, 93, 131
Missouri Avenue, 92, 93, 132
Monroe, James, 75, 93

Monroe Street, 93, 106, 107
Montana Avenue, 93, 132
Montello Avenue, 106
Montgomery Blair Portal, 98
Montgomery Street, 103
Moore, Joseph West, 38
Morgan, Daniel, 93
Morgan Avenue, 101
Morgan Street, 93
Morningside Drive, 93
Morris, Robert, 94
Morris Place, Road, 94
Morris Street, 104
Morse, Samuel F. B., 72, 94
Morse Street, 94, 106
Morton Street, 106
Mount Olivet Road, 94
Mount Pleasant, 94
Mount Pleasant Street, 94
Mozart, Wolfgang Amadeus, 94
Mozart Place, 94
Murdock, George, 95
Murdock, William, 95
Murdock, William D.C., 95
Murdock Mill, 95
Murdock Mill Road, 95
Murdock Place, 106
Myrtle Street, 47

N

National Arboretum, 36, 39, 42
National Cathedral, 20, 73
Navy Yard, 20, 21
Neal Street, 106
Nebraska Avenue, 22, 49, 65, 83, 95, 106, 132
Nelson, Thomas, 95
Nelson Place, 95
Nevada Avenue, 95, 132
New Hampshire Avenue, 96, 106, 132
New Jersey Avenue, 78, 96, 132
New Mexico Avenue, 96, 106, 132
New Orleans, Louisiana, 25
New York Avenue, 132
Newark, New Jersey, 25
Newark Street, 25, 95, 107
Newcomb, Simon, 95
Newcomb Street, 95
Newport, Rhode Island, 25
Newport Place, 25, 96
Newton, John, 96
Newton Place, Street, 96, 106
Nichols, Charles Henry, 88
Nichols Avenue, 88
Nicholson, Sir Francis, 97
Nicholson Avenue, Street, 97, 107
North Capitol Street, 97
North Carolina Avenue, 98, 132
North Dakota Avenue, 98, 132
North Portal Drive, 98
Nourse, Joseph, 20

Index

O

O Street, 106
O'Brien, Robert Lincoln, 27
O'Neal, Margaret "Peggy," 39
Oak Avenue, 103
Oak Court, 106
Oak Drive, Street, 98
Oak Hill Cemetery, 59
Oak Street, 47, 106
Oakdale Place, 47, 98
Oakview Terrace, 47
Oakwood Street, Terrace, 47, 98
Oates Street, 106
Observatory Circle, Lane, Place, 98, 99, 102
Officer Kevin J. Welsh Memorial Bridge, 154
Oglethorpe, James, 100
Oglethorpe Street, 100, 107
Ohio Drive, 17, 100, 132
Oklahoma Avenue, 98, 132
Old Post Office, 92
Old Post Road, 1
Olive Street, 47, 100
Omaha Street, 107
Oneida Place, Street, 100
Ontario Avenue, 106
Ontario Place, 106
Ontario Road, 100, 106
Ontario Street, 106
Orange Street, 47, 100
Orchid Street, 47, 100
Ord, Edward Otho Cresap, 101
Ord Street, 101
Ordway Street, 101, 107
Oregon Avenue, 101, 107, 132
Orleans Place, 25, 109
Otis, James, 109
Otis Place, Street, 107, 109

P

P Street, 107
P Street Bridge, 158
Pacific Circle, 38
Paine, Thomas, 71
Palisade Lane, 109
Paper Mill, 109
Papermill Court, 109
Park Road, 107, 109
Park Street, 107
Parkside Drive, 109
Patent Office, 92
Patricia Roberts Harris Drive, 109
Patrick, Marsena Rudolph, 109
Patrick Circle, 110
Patterson, Edgar, 109
Payne, William H.F., 111
Payne Terrace, 111
Peabody Street, 108
Pecan Street, 47, 111
Peirce, Abner C., 114
Peirce, Isaac, 114
Peirce, Joshua, 80
Pemberton, John C., 15
Penn, William, 111
Penn Street, 107, 111
Pennsylvania Avenue, 17, 69, 111, 132
Pension Building, 90
Perry, Oliver Hazard, 51, 113
Perry Place, Street, 107, 113
Pershing, John Joseph, 113
Pershing Drive, 113
Peterson, August, 23
Phelps, John S., 113
Phelps Place, 113
Philadelphia Street, 106
Pickerell, Adolphus, 137
Pierce, Franklin, 32, 51, 114
Pierce Mill Road, 114
Pierce Street, 114
Pierrepont Place, 106
Piney Branch Parkway, Road, 101, 114
Piney Branch Valley, 158
Pleasant Street, 58
Polk, James K., 8, 115
Polk Street, 115
Pomeroy Street, 108
Pond Street, 115
Pope, John Russell, 115
Pope Street, 115
Poplar Street, 47, 101, 115
Port Eads, Louisiana, 39
Porter, David, 12, 115
Porter, David Jr., 115
Porter, William D., 115
Porter Place, Street, 115
Portland, Oregon, 25
Portland Street, 25, 116
Potomac Avenue, 20, 59, 107, 116
Potomac River, 11, 116
Powhatan, 116
Powhatan Place, 116
President's Square, 33, 75
Presidents, 116
Pretty Prospect, 99, 118
Primrose Road, 47, 117
Prince George's County, Maryland, 15
Princeton, New Jersey, 25
Princeton Place, 25, 118, 138
Princeton Street, 104, 138
Procter, A. Phimister, 109, 112, 113
Prospect Street, 102, 118
Providence, Rhode Island, 25
Providence Street, 25, 106, 118
Puerto Rico, 118
Puerto Rico Avenue, 118

Q

Q Street, 103, 107
Quakenbos Street, 50, 107

Index

Quarles, William A., 118
Quarles Street, 118
Quarry Place, 118
Quebec, Canada, 25
Quebec Place, Street, 25, 107, 119
Queen Anne of England, 119
Queen Anne's County, Maryland, 119
Queen Anne's Lane, 119
Queen Street, 102, 107
Quincy, Josiah, 119
Quincy Place, 107
Quincy Street, 107, 119

R

R Street, 107
Railroad Avenue, 119
Raleigh, North Carolina, 25
Raleigh, Sir Walter, 119
Raleigh Street, 25, 119
Randolph, Edmund, 119
Randolph Place, 119
Randolph Street, 107, 108, 119
Raoul Wallenberg Place, 120
Raum Street, 107
Ray, James Earl, 88
Ream, Vinnie, 45
Redbud Lane, 47, 120
Redwood Drive, Terrace, 47
Reno, Jesse Lee, 120
Reno Road, 107, 120
Renwick, James, 30
Reservoir Road, Street, 121
Rhode Island Avenue, 122, 132
Rhododendron Valley Road, 47, 122
Richardson, William A., 122
Richardson Place, 122
Richmond Street, 102, 107, 108
Ridge Road, 49
Ridge Street, 104, 106, 107
Riggs, George Washington, 122
Riggs, George Washington Jr., 122
Riggs Bank, 22, 122
Riggs Mill, 122
Riggs Place, 107, 122
Riggs Road, 122
Rittenhouse, David, 122
Rittenhouse Street, 107, 122
River Road, 123
Roanoke Street, 103
Rochambeau, Jean Baptiste Donatien de Vimeur, comte de, 77, 158
Rochambeau Bridge, 154, 158
Rock Creek, 16, 40, 70, 92
Rock Creek and Potomac Parkway, 123
Rock Creek Church, 123
Rock Creek Church Road, 51, 123
Rock Creek Drive, 107
Rock Creek Ford Road, 123

Rock Creek Park, 10, 73, 80, 83, 92
Rock Hill, 64
Rock of Dumbarton, 36
Rodman, Thomas J., 123
Rodman Street, 107, 123
Romula, Carlos P., 10, 30
Roosevelt, Franklin D., 29
Roosevelt, Theodore, 22
Rosedale Street, 107, 124
Rosemount Avenue, 124
Ross, J.W., 124
Ross Drive, Place, 124
Rusk Street, 106

S

S Street, 102, 107
Salem, Virginia, 25
Salem Lane, 25, 124
Saratoga, New York, 25
Saratoga Avenue, 25, 124
Saul, John, 47
Savannah, Georgia, 25
Savannah Place, Street, Terrace, 25, 104, 107, 108, 124
Scott, Gustavus, 109
Scott, Winfield, 9, 51, 125
Scott Avenue, 106
Scott Circle, Place, Circle Drive, 125
Scott Circle, 10, 30
Seaton, William, 54, 125
Seaton Place, 107, 125
Seaton Street, 107, 125
Sedgwick, John, 125
Sedgwick Street, 107, 125
Service Road, 126
Seward, William H., 9
Shakespeare, William, 5, 7
Shepherd, Alexander Robey, 112, 126
Shepherd Road, Street, 107, 108, 126
Sheridan, Philip H., 128
Sheridan Circle, Drive, Road, Street, 107, 128
Sherman, John, 128
Sherman, William Tecumsah, 81, 91, 128
Sherman Avenue, 128
Sherman Circle, 128
Sherman Street, 104
Shoreham Bridge, 158
Sidney, Sir Philip, 5
Silver Spring, Maryland, 39
Smithsonian Institution, 76
South Capitol Street, 20, 21, 59, 88, 129
South Capitol Street Bridge, 110
South Carolina Avenue, 36, 129, 132
South Dakota Avenue, 130, 132

Index

South Portal Drive, 130
Southeast-Southwest Freeway, 130
Southern Avenue, 130
Spring Hill, 52
Spring of Freedom Street, 130
Spruce Drive, Street, 47, 130
St. James Creek, 20, 29
St. John's in Georgetown, 96
St. Louis, Missouri, 25
St. Louis Street, 25, 124
Stanton, Edwin M., 30, 51, 70, 130
Stanton Road, Terrace, 130
State Streets, 131
Staughton Street, 102
Steuben Street, 103
Stevens, Isaac Ingalls, 50
Stevens, Thaddeus, 132
Stevens Road, 132
Stoddert, Benjamin, 3, 118, 132
Stoddert Place, 132
Stowe, Harriet Beecher, 11
Stratford-upon-Avon, England, 8
Stuyvesant, Peter, 133
Stuyvesant Place, 133
Sudbury, Massachusetts, 25
Sudbury Lane, Place, Road, 25, 133
Suit, Samuel Taylor, 133
Suitland Parkway, 49, 133
Sullivan, Anne, 22
Summit Place, 108
Summit Street, 108
Sumner, Charles, 133
Sumner Road, 133
Sumner Street, 104
Susquehannah Street, 108
Sycamore Drive, Place, Street, 47, 133

T

T Street, 108
Tahoe Street, 102
Takoma Park, 39
Tamarach Street, 134
Tayloe, Benjamin Ogle, 75
Taylor, Zachary, 134
Taylor Street, 108, 134
Tenley Circle, 134
Tenley Place, 108
Tennally, John, 134
Tennessee Avenue, 73, 132, 134
Tennyson, Alfred Lord, 135
Tennyson Street, 135
Texas Avenue, 132, 135
Thayer, John M., 135
Thayer Street, 135
Theocritus, 5
Theodore Roosevelt Memorial Bridge, 159
Thomas, George H., 135
Thomas Circle, Street, 108, 135

Thoreau, Henry David, 43
Thornton, William, 136
Thornton Place, 136
Tiber Creek, 20, 29
Tilden, Samuel Jones, 62, 136
Tilden Place, Street, 108, 136
Timberlake, John B., 39
Titanic, 8
Todd Place, 108, 137
Totten, Joseph Gilbert, 51
Tracy, Benjamin F., 137
Tracy Place, 108, 137
Treasury Building, 92
Trenton, New Jersey, 26
Trenton Place, Street, 26, 108, 137
Trinidad Avenue, 137
Trinidad Street, 101
Truman, Harry S, 85
Trumbull Street, 102
Tuckerman, Walter, 137
Tuckerman Street, 108, 137
Tudor Place, 137
Tulip Street, 137
Tunlaw Farm, 137
Tunlaw Road, 106, 137
Twining, 4
Tyler, John, 61

U

U Street, 108
Umatilla Street, 102
Underwood, Joseph R., 138
Underwood Place, Street, 138
Union Arch Bridge, 90
Uniontown, 4
Universities, 138
University Avenue, Place, Terrace, 138
Upshur, Abel P., 138
Upshur Street, 108, 138
Upton, Emory, 139
Upton Street, 108, 139
USS *Benton*, 12
Utah Avenue, 132, 139
Utica Street, 108

V

V Street, 108
Vallejo Street, 16, 102, 108
Van Buren, Martin, 14, 35, 71, 139
Van Buren Street, 139
Van Ness, John Peter, 18, 59, 140
Van Ness, Marcia Burnes, 18, 59, 140
Van Ness Street, 108, 140
Varnum, James Mitchell, 140
Varnum Place, Street, 108, 140
Verbena Street, 48, 140
Vermillion Street, 103
Vermont Avenue, 132, 140

Index

Vernon Avenue, 108
Verplanck Place, 108
Vietnam Veteran's Memorial, 8
Vine Street, 48, 140
Virginia Avenue, 132, 140
Volta Place, 142

W

W Street, 108
Wabash Street, 103
Wade, Benjamin F., 142
Wade Road, 142
Wahunsonacock, 116
Waldorf-Astoria Hotel, 7
Wallace Street, 101
Wallenberg, Raoul, 120
Walnut Avenue, Street, 48, 108, 142
Walter, Thomas U., 90, 142
Walter Street, 142
Ward, Artemas, 73, 142
Ward Circle, Place, 142
Warder Avenue, 108
Warder Street, 108
Warren Place, 108
Warren Street, 108
Warwickshire, England, 4
Washington, George, 9, 15, 41, 60, 73, 74, 77, 143
Washington Avenue, 20, 132, 143
Washington Canal, 20, 29
Washington Channel, 144
Washington Circle, 143
Washington Monument, 30, 42, 92
Water Street, 144
Waterside Drive, 108, 144
Wayne, Anthony, 144
Wayne Place, 144
Webster, Daniel, 144
Webster Street, 108, 144
Welling Place, 103
Wesley, Charles, 145
Wesley, John, 144
Wesley Circle, Place, 144
West Basin Drive, 145
West Beach Drive, Terrace, 145
West Executive Avenue, 145
West Virginia Avenue, 132, 146
Western Avenue, 145
Western Street, 108
Westminster Abbey, 146
Westminster Street, 146
Wetherell, C.S., 143
Wheeler, Joseph, 146
Wheeler Road, 146
White House, 26, 39, 42, 111
Whitehaven Parkway, Street, 146
Whitehurst Freeway, 146
Whitney, Asa, 109
Whitney, Catherine M., 109
Whitney Avenue, 107, 109
Whitney M. Young Jr. Memorial Bridge, 159
Whittier Street, 108
Willard, Henry, 147
Willard Street, 147
William Howard Taft Bridge, 159
Williams, Elie, 109
Williamsburg, Virginia, 26
Williamsburg Lane, 26, 147
Willow Avenue, 108
Willow Street, 48, 108, 147
Wilmington, Delaware, 26
Wilmington Place, 26, 147
Wilmington Street, 103
Wilson, Woodrow, 22, 159
Wilson Street, 108
Windom, William, 148
Windom Place, 108, 148
Wisconsin Avenue, 49, 102, 132, 148
Woodley Place, Road, 101, 103, 108, 148
Woodley Street, 108
Woodrow Wilson Bridge, 159
Wright, Orville, 23
Wright, Wilbur, 23
Wright's Road, 102
Wyoming Avenue, 132, 149

X

X Street, 1
Xenia, Ohio, 26
Xenia Street, 26, 108

Y

Y Street, 1, 108
Yale Street, 44, 104, 138
Yorktown, New York, 26
Yorktown Road, 26, 149
Young, Joseph, 57
Young, Notley, 96
Yuma, Arizona, 26
Yuma Street, 26, 103, 149

Z

Z Street, 1
Zane Street, 23
Zanesville Street, 104